JUNIATA COLLEGE
LIBRARY
Mr. and Mrs. David O. Harbaugh
Memorial Book Fund

THE YOUNG OXFORD BOOK OF THE

Prehistoric World

THE YOUNG OXFORD BOOK OF THE

Prehistoric World

Jill Bailey   Tony Seddon

Oxford University Press

Contents

Oxford University Press

Oxford New York
Athens Auckland Bangkok Bombay
Calcutta Cape Town Dar es Salaam Delhi
Florence Hong Kong Istanbul Karachi
Kuala Lumpur Madras Madrid Melbourne
Mexico City Nairobi Paris Singapore
Taipei Tokyo Toronto

and associated companies in
Berlin Ibadan

Published by Oxford University Press, Inc.
200 Madison Avenue
New York, New York 10016

Oxford is a registered trademark of Oxford University Press

Library of Congress Cataloging-in-Publication Data

Bailey, Jill.
Prehistoric World / Jill Bailey, Tony Seddon.
p. cm.—(Young Oxford books)
Includes index.
1. Historical geology—Juvenile literature.
[1. Historical geology.]
I. Seddon, Tony. II. Title. III. Series.
QE28.3.B33 1995
560—dc20 95-7014
CIP
AC

ISBN 0-19-521163-4 (trade ed.)
ISBN 0-19-521162-6 (lib. ed.)

9 8 7 6 5 4 3 2 1

Printed in Spain by
Mateu Cromo. S.A. Pinto—Madrid

Front cover *Uintatherium*, a large plant-eating mammal of the Eocene (see page 122).

1

The Evolving Planet

2

The History of Life

CONTENTS

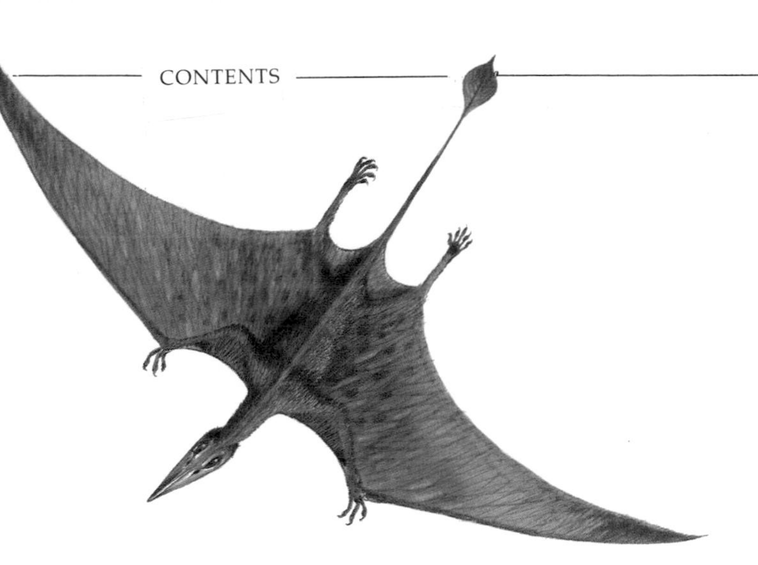

Introduction

This is the story of how Earth developed from a hot, fiery planet to a world of blue oceans and green continents. A hostile land of bare rock, volcanoes, and unpleasant gases was transformed by life itself – the rock became covered in fertile soil, and the atmosphere and the waters of Earth were enriched with oxygen.

Earth today is teeming with life – the home, according to one estimate, to some 3,000 million million million million million living things. There are probably between 2 and 30 million different kinds of plants and animals, and every year about 10,000 new species of animals and 5,000 new species of plants are discovered.

Even so, all these living species represent just a fraction of those that have existed throughout Earth's history. One guess puts the figure at 500 million species, but what is certain is that most of the plants and animals that have existed on Earth have disappeared forever. Bursts of evolution have been followed by mysterious mass extinctions, when large numbers of species suddenly died out. In turn, their places were taken by new and different life forms, some of which became ever more complex until a creature evolved that was to transform Earth more dramatically than any that had come before it: the human being. This book is the story of Earth and the creatures that live and have lived upon it: the story of their lives and deaths, and of their contribution to the rich and complex world that is Earth today.

Tony Seddon
Jill Bailey

THE EVOLVING PLANET

During the 4.6-billion-year history of Earth, millions of plant and animal species have come and gone; vast mountain ranges have risen and been worn away; and continents have broken up and drifted across the globe, then collided again to form new landmasses. How do we know? Despite these great upheavals, a surprising amount of this history remains written in the rocks that survive today, in the fossils they contain, and in the bodies of living things. This record is fragmented – we are given only glimpses at infrequent intervals; vital chapters are missing from the story. Yet the story itself is as gripping as any detective novel.

THE ORIGIN OF THE EARTH

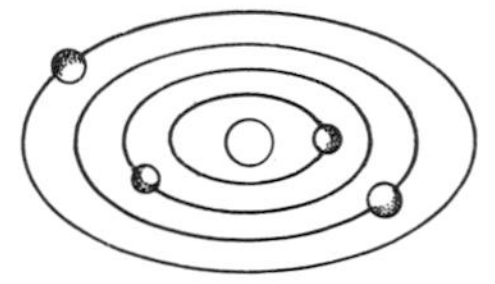

There are about 100 billion stars in a galaxy, and 100,000 million galaxies in the universe. If you could travel from Earth to the edge of the universe, it would take you more than 15 billion years traveling at the speed of light, which is 186,000 miles per second. But where did all this material come from? How was the universe formed?

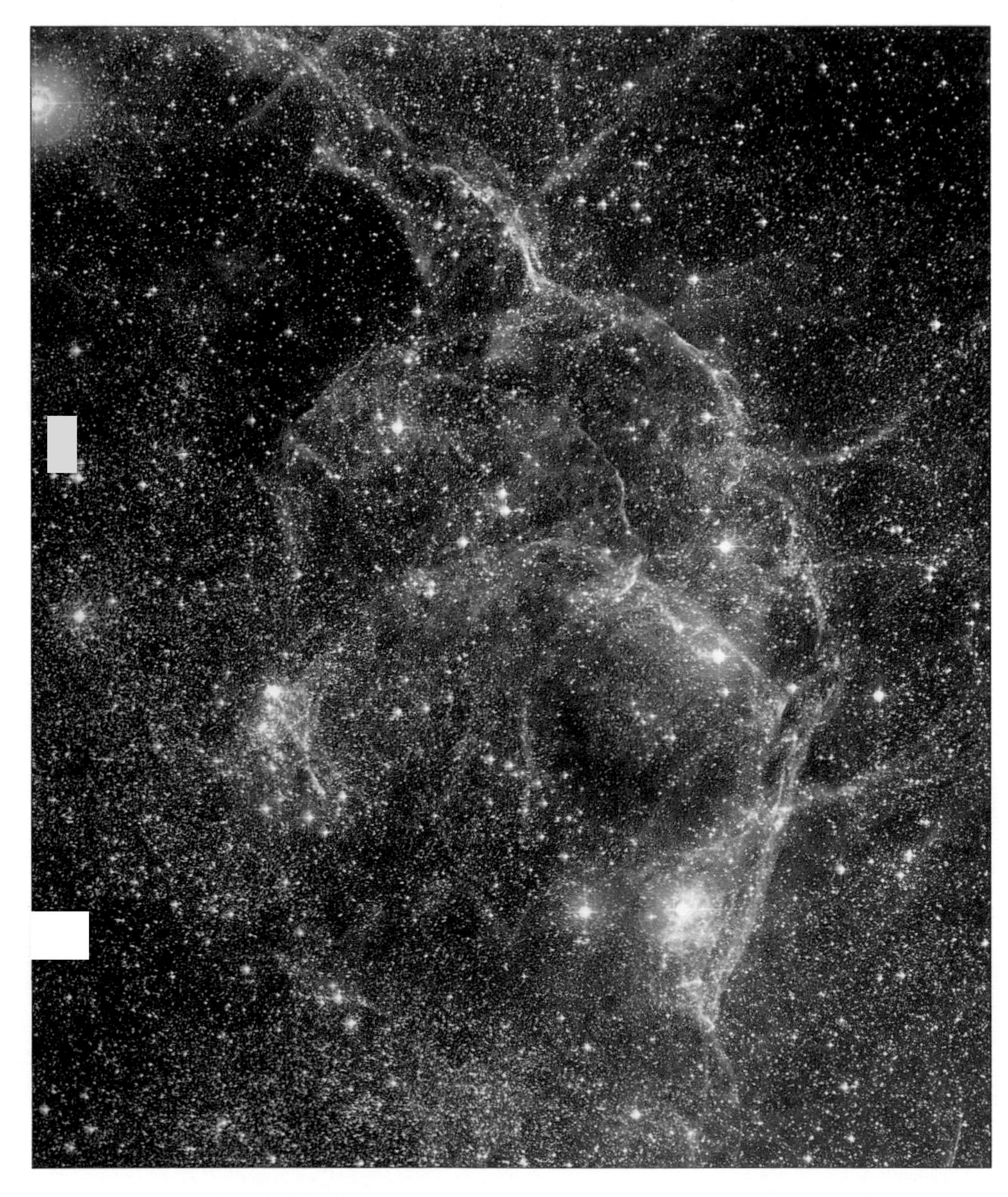

The best theory we have today for the formation of the universe suggests that it all began with a big bang. An incredibly hot fireball, with a temperature of thousands of millions of degrees, exploded and sent energy and particles of matter speeding out into space.

Matter is composed of particles called atoms. These are the smallest pieces of matter that can take part in a chemical reaction. They are made up of even smaller particles. There are many different kinds of atoms, which are called elements. Each element has atoms of a different size and weight from those of other elements, and behaves differently in chemical reactions. Everything in the universe, from the largest galaxies to the tiniest living things, is made from chemical elements.

After the big bang

At the huge temperatures of the big bang fireball, the tiny particles that make up matter had too much energy to combine into atoms. After about a million years, when the universe had cooled to 7,200 °F, atoms began to form. The first elements formed were the lightest – helium and hydrogen. As the universe cooled further, heavier elements were formed. New atoms and elements are still forming today inside stars such as the Sun, where temperatures are extremely high.

As the universe cooled still further, the newly formed atoms came together to form great clouds of dust and gas. As dust particles collided and merged together, gravity began to work, attracting small objects toward larger ones. Gradually stars, planets, and galaxies evolved.

The expanding universe

The big bang was so powerful that the matter in the universe was flung out into space. The universe is still expanding today. We know this because distant galaxies are still speeding away from us as the space between them expands. This is evidence that the galaxies were once much closer together.

◀ Astronomers think the world began with a big bang. A huge fireball exploded, flinging matter and energy out into space, where they condensed to form billions of stars clustered into galaxies.

Microwaves from the past

Assuming the universe was formed by a "hot" big bang, starting with a great fireball, scientists have calculated how much it should have cooled by now. They estimate that the space between the galaxies should have a background temperature of about -450 °F. Scientists can take the temperature of the universe by measuring the microwave (heat) radiation in space. This does indeed give a temperature of -450 °F.

How old is the universe?

Astronomers use measures like the size and brightness of galaxies, and the color of the light coming from them, to figure out how far away they are. If the big bang theory is true, then in the beginning all the galaxies were squeezed into one dense, hot fireball. So if you divide the distance from one galaxy to the next by the speed at which they are moving apart, you can find

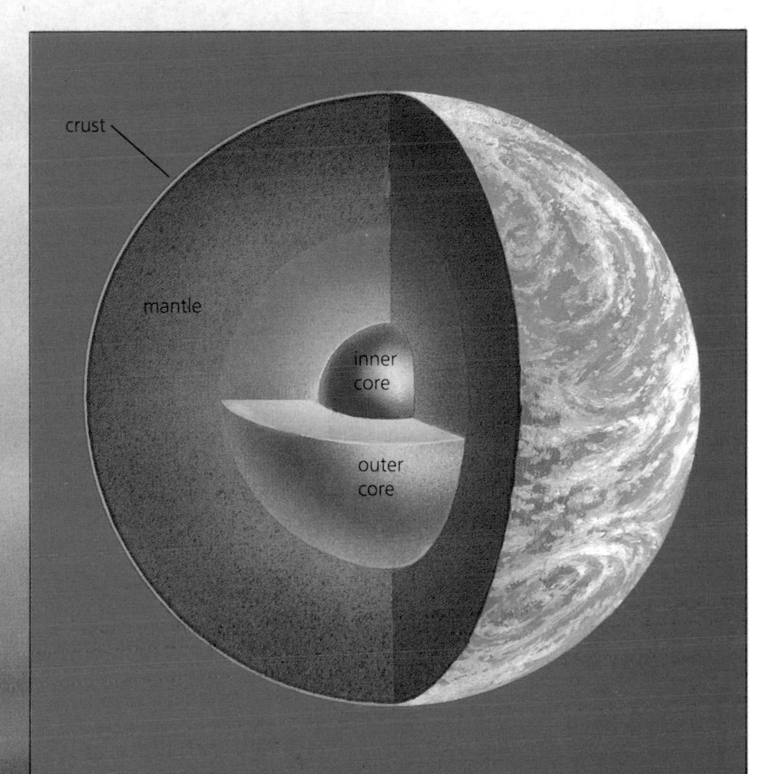

► Earth has a molten core rich in iron and nickel. The crust, which is of lighter material, floats on the partly melted rocks of the mantle.

▼ Nobody knows for certain how the Solar System was formed. The main theory is that the Sun and planets condensed from a swirling cloud of gas and dust. Denser parts of the cloud attracted more matter by the pull of their gravity, forming the Sun and the planets.

out when they were together. This is the age of the universe. These methods are not very accurate, but we think the universe is about 12-20 billion years old.

Formation of the Solar System

The galaxies were probably formed about 1-2 billion years after the big bang, and the Solar System came into being 8 billion years later. Matter was not evenly spread in space. Denser areas attracted more dust and gas owing to the pull of gravity, so they grew faster and faster, until they formed great whirling clouds of dust and gas called nebulas.

One particular nebula – the Solar nebula – condensed to form the Sun. Other parts of the cloud condensed to form the planets (including Earth), which were held in their orbital paths by the pull of the Sun's gravity. As gravity pulled the contents of the Sun closer and closer together, the Sun became smaller and denser. The immense pressure at the Sun's core created a lot of heat and allowed the fusion reactions producing new atoms to go even faster, generating still more heat.

Getting ready for life

Something similar was happening to Earth, but on a smaller scale. So much heat was produced by its collapsing core, and by nuclear reactions and the decay of radioactive substances inside the earth, that the rocks melted. The lighter crust material – rich in the glasslike mineral silica – separated from the denser iron and nickel in the earth's core. After about a billion years, when Earth had cooled still further, the hard outer crust of rock had formed.

As Earth cooled, gases were expelled from its core, usually through erupting volcanoes. Light gases such as hydrogen and helium were mostly lost to space. However, the pull of Earth's gravity was strong enough to hold on to the heavier gases, and they formed the atmosphere. Some of the water vapor condensed to form the oceans. Earth was now ready to support life.

The death and birth of rocks

The land is formed from solid rocks, often covered by soil and vegetation. But where did these rocks come from? New rocks are formed from materials produced deep inside the earth. Way down in the earth's crust the temperature is much hotter than at the surface, and the rocks are under great pressure. The heat and pressure make them become bendy, and even liquid. Where there is a weakness in the earth's crust, the molten rock, called magma, wells up to the surface. It flows out onto the surface as lava. As the lava cools, it becomes solid rock.

Explosions and fire fountains

The birth of rocks can be a very violent affair, or it can be quiet and unspectacular. There are many different kinds of magma, and they produce different rocks. Basalt-forming magma flows easily and quickly onto the surface, where it spreads out into wide sheets that cool quickly. Sometimes it bursts out of a volcano in a red-hot "fire fountain" as the pressure is released.

Other magmas are much thicker, with a consistency more like molasses. It is difficult for trapped gases to bubble up through thick magma. Think how easily air bubbles out of boiling water, and how slowly it bubbles when you heat up something thicker, like pudding. As this thicker kind of magma gets near the surface, the pressure on it decreases and the dissolved gases want to expand, but they cannot. When the magma finally escapes, the gases expand so rapidly that they cause a great explosion, shooting out lava, rocks, and ash. Mount Pelée on the Caribbean island of Martinique erupted like this in 1902. This catastrophic eruption destroyed the port of Saint Pierre, killing about 30,000 people.

Growing crystals

Rocks formed from cooling lava are called volcanic rocks, or igneous rocks. As the lava cools, the minerals in the molten rock become solid crystals. If the lava cools quickly, the crystals do not have much time to grow, so they are very small. This happens in basalt. Sometimes the lava cools so fast that it produces a smooth, glassy rock with no crystals, like obsidian. This is likely to happen if the lava pours out underwater, or if small blobs of lava are flung high into the cool sky.

 Lava streams down the flanks of Kilauea volcano. When lava reaches the earth's surface, it cools, forming new rock.

CLUES TO THE PAST

Crystal size in volcanic rocks tells us how fast the lava cooled, and whether it was near the surface or not. This is a piece of granite viewed with polarizing light under a microscope. The different crystals show up in different colors.

Gneiss is a metamorphic rock formed from a sedimentary rock that was changed by heat and pressure. The patterns of the different-colored bands in this block of gneiss show the direction in which the layers of rocks were squeezed by movements of the earth's crust, clues to events that happened 3.5 billion years ago.

Folds and faults (breaks) in the rocks tell us about the direction of great pressures in the earth's crust long ago. These folds were formed by mountain-building movements that began 26 million years ago. Here, great forces have bent and folded layers of sedimentary rocks.

The magma does not always get as far as the surface. It may cool much more slowly, deeper in the crust, and beautiful large crystals form. Granite is made in this way. The size of the crystals in some pebbles can give clues to how the rock was formed millions of years ago.

Sediment sandwiches

Not all rocks look like the volcanic rocks granite and basalt. Many cliffs look as if they are made up of layers like a pile of sandwiches. They have been formed from other rocks that have been worn down by wind, rain, and rivers to form sediments that are washed into lakes and seas. These sediments pile up on the bottom of the lake or sea, often hundreds or even thousands of feet thick. The sediments at the bottom are under great pressure and this squeezes out the water. Minerals deposited from the water help to bind the sediments together to form new rocks called sedimentary rocks.

Both volcanic and sedimentary rocks can be pushed up by movements of the earth's crust to form new mountain ranges. The forces involved in mountain building are tremendous, and the rocks may become very hot, or be squeezed very hard. This can alter them, changing one mineral into another, flattening the crystals and rearranging them to form different kinds of rocks. Rocks that have been formed by changing other rocks in this way are called metamorphic rocks.

Rocks that come and go

Nothing seems more solid and everlasting than a great mountain. But this is just an illusion. On the geological time scale, where we are thinking in terms of millions and even hundreds of millions of years, mountains come and go.

As soon as a rock is exposed to the atmosphere, it begins to break down. If you look at a piece of recently broken rock or pebble, you will see that the newly exposed rock surface is often quite a different

▶ Erosion and weathering in the canyons of Cedar Breaks, Utah. The canyons were formed by the erosive power of a river, cutting into the sedimentary rocks as they were uplifted by earth movements. The exposed rock faces are being weathered, and the debris forms scree slopes. Hard bands of rock stand out from the screes, forming the rims of the canyons.

color from the old exposed surface. This is due to the effects of oxygen in the atmosphere, and often to rain as well. Chemical reactions have taken place that change the nature of the rock surface.

In time, these reactions loosen the minerals in the rock, so it starts to crumble. Water seeps into tiny cracks in the rock. When water freezes, it expands, forcing the rock apart. When the ice melts, the rock will fall to pieces. Rain soon washes the loose particles away. These processes are called weathering.

Water, the destroyer

The loosened rocks are swept into rivers, where they tumble along the riverbed, scouring away the rocks below, until the surviving fragments finally come to rest in a lake or in the sea. Frozen water (ice) is even more destructive. Glaciers and ice caps have large and small rocks trapped in their sides and bellies, which scrape deep grooves in the rocks below. Rock fragments that fall onto the top of the ice may be carried for hundreds of miles.

Sculptures of the wind

Even the wind can erode rocks. In deserts in particular, the wind carries millions of grains of sand. Sand is made mostly of quartz, an extremely hard mineral. This sandblasts the rocks, producing more and more sand.

The wind may pile up the sand into dunes. Each gust of wind builds another layer of sand grains onto the dunes. The angle and direction of slope of these layers are a clue to the direction and strength of the wind that piled them up.

◀ Top: The Hoodoos, Alberta, Canada. Rain and sandblasting by the wind have worn away softer rocks faster than harder rocks, leaving behind rock formations with strange shapes.

Center: The Muir Glacier in Alaska. The scouring action of the glacier, and of the stones embedded in its base and sides, erodes the valley sides and floors, forming stripes of rock debris (moraines) on the ice. Moraines from adjacent glaciers merge where the glaciers meet.

Bottom: Glaciers gouge out deep U-shaped valleys. In Nant Ffrancon, Wales, the glaciers disappeared long ago, leaving a valley far too wide for the small stream that now occupies it. The small lake in the foreground is dammed up by a bar of extra hard rock.

The Record in the Rocks

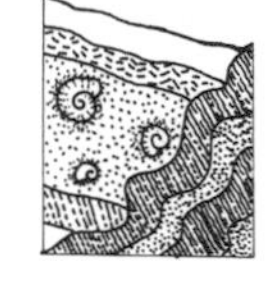

Earth's history is written in the rocks. But the rocks are like the pages of a torn book – you must put them in the right order if you are to understand the story. In most places, the oldest rocks are at the bottom and the youngest ones are at the top. But rocks may not stay in the same place forever.

The tremendous forces that push up new mountains can tilt, fold, and crumple the rocks. Sometimes the folds break off or curl over so that the oldest rocks, which were once at the bottom, are now at the top.

Rocks may break and slide past one another, so that old rocks lie alongside younger ones. Or volcanic rocks may be forced up from deep in the earth's crust until they push through younger rocks. If the volcanic rock is found to have pushed through several layers of other rocks, it must have been formed later.

Erosion may wear down the rocks, producing patterns in which the oldest rocks are surrounded by younger ones. This can happen if the top of a fold is worn away, exposing the older rocks in the middle. New sediments may then be deposited on the old erosion surface.

Numbering the pages

To sort out the true story you need clues to tell you in which order the rocks were formed. Fossils – the remains of plants and animals preserved in the rocks – are very good clues. Over millions of years, the earth's climate has changed, and so have the plants and animals that live on the planet.

The most useful clues come from hard-shelled animals and, for more recent rocks, from vertebrates (animals with backbones). Their hard parts are more likely to be preserved than the soft tissues. Mollusks and other shellfish are the most familiar fossils, but tiny one-celled animals, too

◀ An unconformity in the rocks tells a story – limestone from the Carboniferous period overlies Silurian rocks that have been folded by earth movements, then eroded to produce the flat surface on which the limestone has been laid down. The uppermost rocks on such an unconformity must be the youngest.

small to see except under a microscope, may also have shells. Rocks that contain exactly the same kinds of animals were probably formed at the same time.

Rocks with built-in clocks

Only sedimentary rocks contain fossils. Different kinds of clues must be found for other types of rocks. The age of a volcanic (igneous) rock can be found by radioisotope dating.

The simplest chemicals in nature are elements. Iron is an element and so is oxygen. These substances are made up of only one kind of atom. But some elements have an unstable form. Their atoms (parent atoms) may break down (decay) to form other kinds of atoms (daughter atoms), giving out radiation as they do so. These unstable forms are called radioisotopes.

Radioisotopes decay at a steady rate. Each kind of radioisotope has its own special rate of decay. If you know the rate at which a particular radioisotope decays, and you know the ratio of daughter isotope to parent isotope in the rock, you can figure out how long the isotope has been decaying. This is the age of the igneous rock.

Radioisotope dating does not work for sedimentary rocks, as it tells you only the age of minerals in the rocks containing the radioisotopes. These minerals came from older rocks that were broken down to form the sediments.

Ancient magnetism

Earth has a molten-liquid core rich in iron. The planet spins around its axis (an imaginary line drawn between the North and South Poles). This makes it behave like a giant magnet. The north and south poles of the magnet are not in quite the same place as the North and South Poles of the earth. For reasons we do not fully understand, the positions of the magnetic north and south poles have swapped from time to time.

Some volcanic rocks have iron-containing minerals that act like tiny magnetic markers. Before the rocks cooled and became solid, the minerals lined up with the earth's north and south magnetic poles. But these poles may not have been in the same place as they are today. Scientists have figured out the positions of the earth's

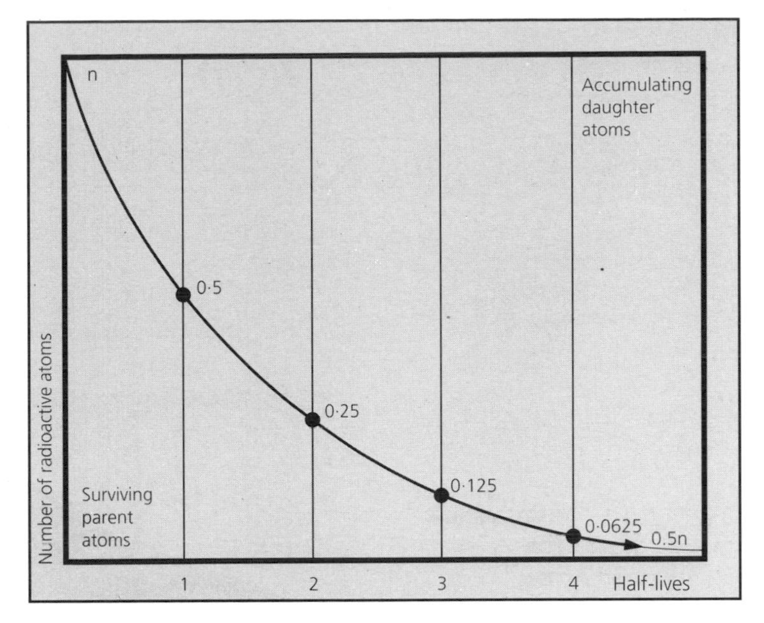

◀ Volcanic rocks can be dated by radioisotope dating. Certain elements in the rocks slowly "decay" from one form to another, giving off radiation in the process. The two forms of the element are called radioisotopes. The element decays at a steady rate, so by measuring the proportion of the two radioisotopes in the rock, you can calculate when it was formed.

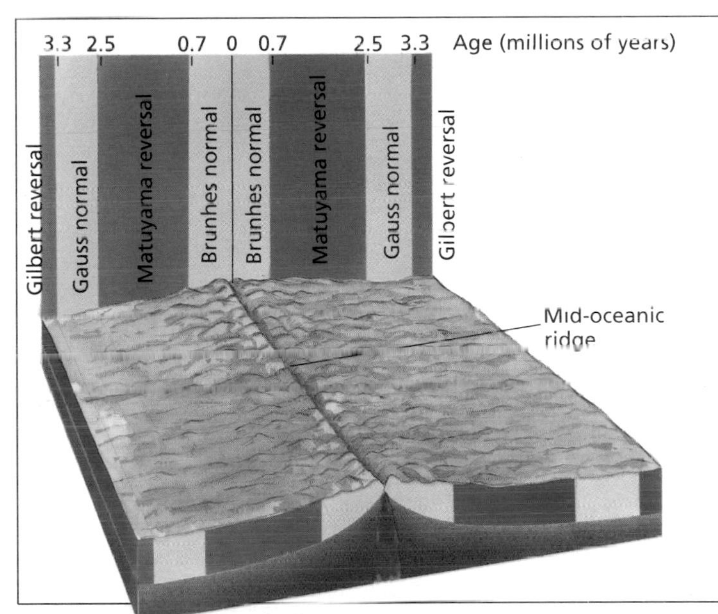

◀ From time to time the earth's magnetic field reverses: The north magnetic pole becomes the south magnetic pole and vice versa. As lava solidifies to form rock, iron-containing minerals line up in the direction of the earth's magnetic field. The diagram shows rocks with different magnetic polarity on either side of a mid-oceanic ridge where new rock is forming. Polarity can help to date the rocks.

◀ A fossil seed fern from the ancient coal swamps. Seed ferns grew only in warm climates. For such soft parts to be preserved, they must have been buried rapidly. The rock is shale, formed from mud deposits, so this area was once a warm swamp forest into which sediment was being washed very fast.

magnetic poles at various times in the past, so from the direction of these tiny magnets, we can sometimes determine the age of rocks.

How old is Earth itself?

The oldest rocks in the world today have been dated to a little less than 4 billion years old. But by the time these rocks were formed, the first rocks had already been worn down to sediments and used to form new rocks. So we cannot calculate the age of Earth by studying the rocks around us today.

However, scientists believe that Earth, the Moon, and the meteorites that occasionally fall to earth were all formed at the same time. The meteorites are fragments of rock from outer space, unchanged since the time they formed, and they have been dated to 4.6 billion years. The oldest rocks on the Moon are also about 4.6 billion years old, so it is likely that Earth is the same age.

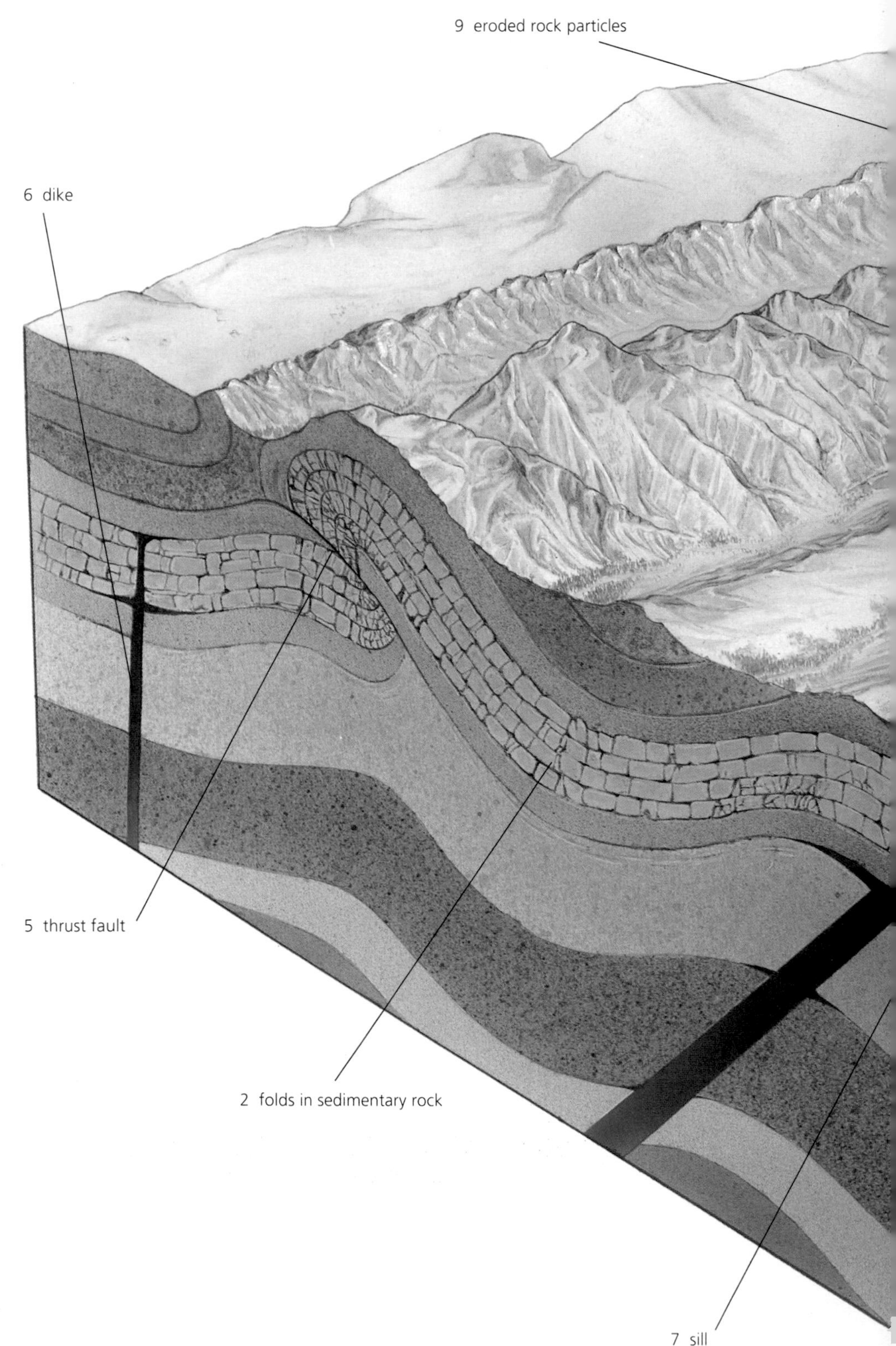

▶ Landscapes are sculpted by the forces of erosion and weathering – by rain, rivers, ice, frost, heat, wind, and chemicals in rainwater. Hard rocks wear down more slowly than soft rocks. A landscape hides a great deal of geological information, which is revealed only when rocks are exposed at the surface in cliffs or quarries.

Sloping layers (1) or curved layers (2) of sedimentary rocks indicate that the rocks have been subject to earth movements, such as folding, since they were laid down.

A discontinuity (3) occurs where older rocks have been eroded, then much younger sedimentary rocks laid down on top. In this case, the discontinuity shows up in the landscape, as the younger rocks are much harder, and after millions of years of erosion they stand up as mountains.

Faults (4 and 5) occur when earth movements are so powerful that the rocks shear and move relative to one another along the shear plane (fault line). In thrust faults (5) one mass of rock is thrust over another, buckling as it moves. This particular fault has brought older rocks to lie on top of younger ones. It has brought harder rocks up against softer ones to form chains of hills.

Intrusions of igneous rocks such as dikes (6) and sills (7) also help to show relationships between rock layers. Where they cut across layers of different ages, the intrusions must be younger than the sedimentary rocks around them. The volcanic rocks being formed by the volcano (8) will create a new discontinuity.

Erosion and weathering processes create new sediments (9), which may in turn form new sedimentary rocks one day.

1 tilted layers of sedimentary rock
3 discontinuity
3 discontinuity
4 fault
8 volcano

WANDERING CONTINENTS

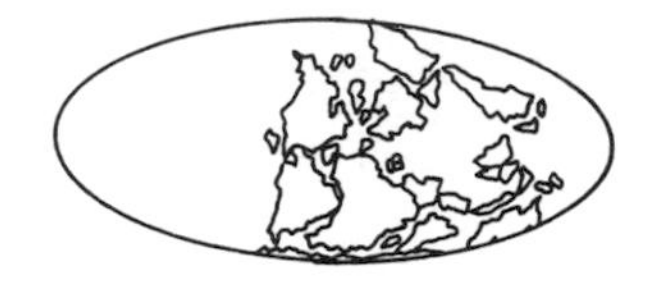

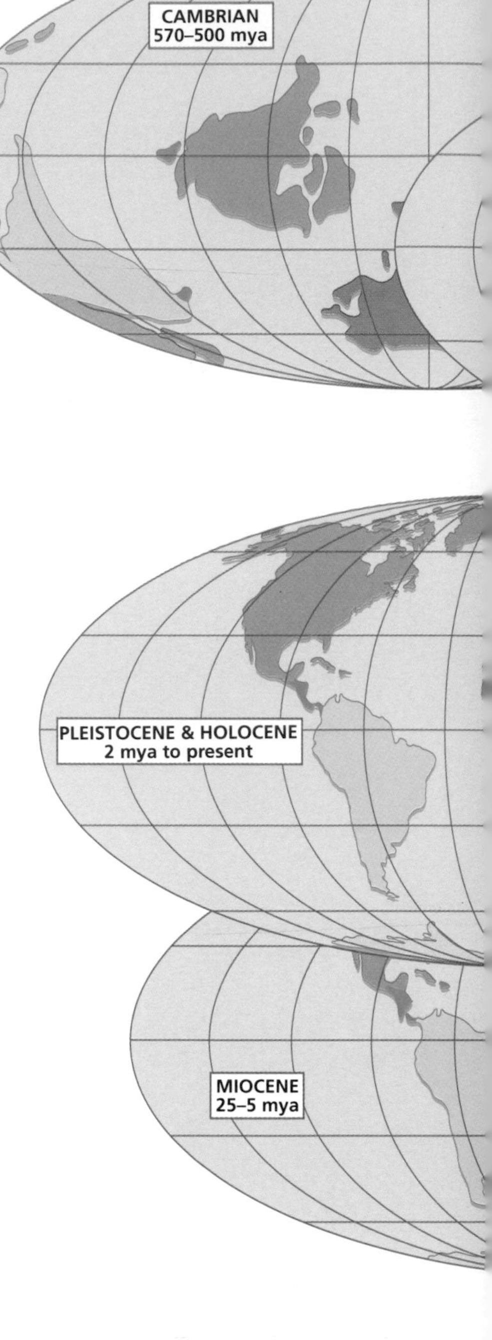

When you "read" the rocks to find out about Earth's past, you discover some astonishing things. In the south of England, for instance, there are sandstones that must have been laid down under hot desert conditions. In Antarctica you find fossils of tropical ferns, and in Africa there is evidence of ice caps. Has the world's climate really been so different in times past?

The earth's climate has changed over geological time: The world has been both warmer and cooler in the past. There have even been a number of ice ages, when ice caps and glaciers spread out from the poles. But the changes have not been so dramatic as the above examples would lead us to believe. The true explanation for these puzzles is even more unexpected: The continents have not always been in the positions they occupy today. Africa was once over the North Pole. India used to be near Africa and actually moved north across the equator until it collided with Asia.

The giant jigsaw puzzle

If you look at the west coast of Africa and the east coast of South America, you can imagine that, if the Atlantic Ocean did not exist, these two continents would fit together very well. The kinds of rocks on the two continents, their ages, and the directions in which they have been folded also fit remarkably well. The best explanation for these "coincidences" is that all these landmasses were once joined together, before the Atlantic Ocean opened up between them.

Probably ever since the earth's crust first cooled to form solid plates, great chunks of the earth's crust have wandered the globe, carrying the continents with them. These movements of the continents are called continental drift. The movements are very slow – at most a few inches a year – but over millions of years they have dramatically altered the map of the world. Sometimes continents have come together to form supercontinents surrounded by giant oceans. These supercontinents in turn have broken up again, forming new seas and lakes, islands and continents.

The idea that the continents might have drifted around the globe was first suggested in 1912 by the German scientist Alfred Wegener. To back up his theory, he looked for evidence in the rocks and in the fossils inside the rocks.

Fossil fingerprints

Over many millions of years the world's climate changes, mountain ranges come and go, and sea levels rise and fall. Some kinds of plants and animals adapt to these changes, others eventually die out. Over geological time new species evolve that are even better suited to the new conditions.

Different changes take place on different continents, and the new plants and animals cannot easily move across the oceans from one continent to another. So each continent and island develops its own special kinds of plants and animals that may later become extinct (die out) if conditions change. The remains of some of these creatures have survived as fossils, and are still there in the rocks for us to find today.

Wegener pointed out that the remains of the freshwater swimming reptile called *Mesosaurus* and the land reptile *Cynognathus* had been found in rocks in South Africa and Brazil, but nowhere else in the world. The remains of a hippolike creature called *Lystrosaurus* are found in rocks from Africa, India, and Antarctica; and fossils of the seed fern *Glossopteris* are found in all the southern continents, but in none of the northern continents. All of these mysteries are explained by the fact that at the time these fossils were formed, the southern continents were all joined together in a supercontinent called Gondwanaland, which was covered in warm, moist forests.

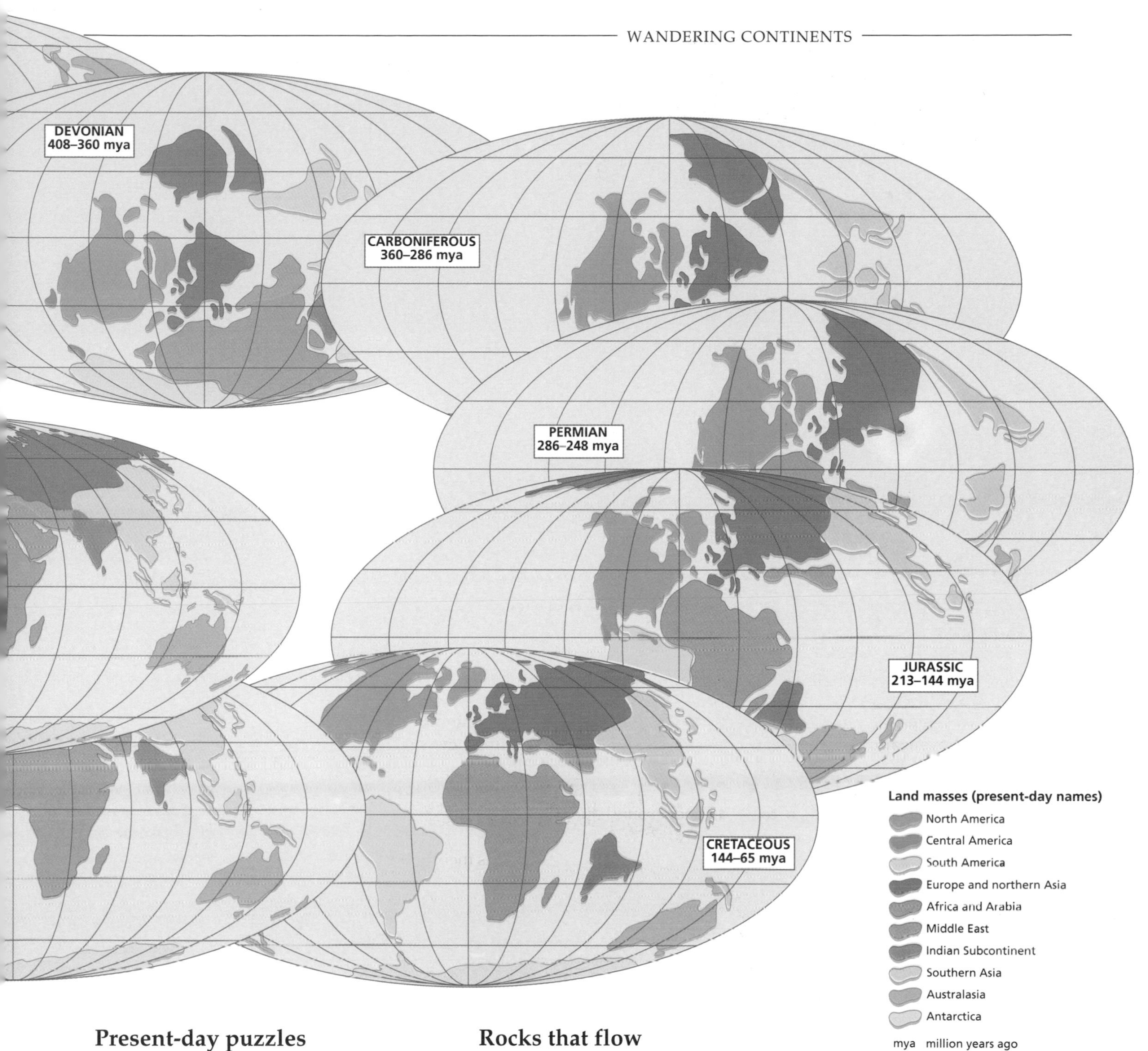

Present-day puzzles

Continental drift also explains the strange distribution of some animals alive today. The island of Madagascar, off the east coast of Africa, is famous for its lemurs. These are monkeylike animals that are thought to resemble primitive monkeys which were widespread some 50 million years ago. Lemurs are not found on mainland Africa. Madagascar became separated from Africa before modern monkeys evolved. On the mainland, these more advanced monkeys became better adapted to finding food and surviving than the lemurs. This competition was too much for the lemurs, which died out.

Rocks that flow

The earth's crust is divided into a number of large rigid sheets of rock called plates. In the mantle (the part of the earth immediately below the crust), the rocks get hotter and hotter as you go downward, until they become bendy, like Plasticine. If they get even hotter, they melt. The lighter plates are really floating on the mantle rocks below.

When a liquid gets hot, it becomes less dense, so the warm liquid rises through the cooler liquid at the surface. The cooler liquid has to flow down to fill the space. The same thing happens in the mantle. The

▲ Continental drift occurs when the continents are riding on huge plates of the earth's crust. These plates are constantly moving very, very slowly, driven by convection currents in the mantle far below. Over millions of years the continents have moved around the globe, passing through different climatic zones as they go. Some plates are growing as new rock forms, pushing over or under others. This, together with the forces of erosion, has changed the shape of the continents. Sometimes they are pushed together to form giant supercontinents. These later break up, forming several smaller continents again.

ISLANDS ON THE MOVE

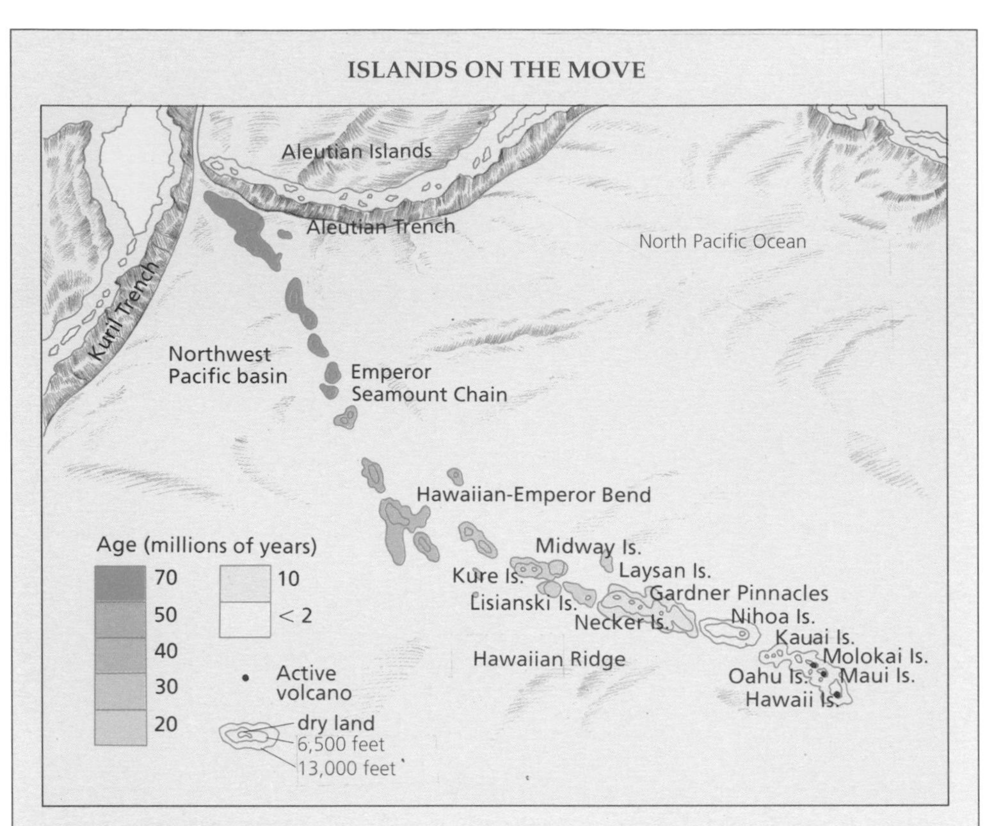

The Hawaiian Islands are part of a long chain of active and extinct volcanoes rising from the floor of the Pacific Ocean. The chain extends northwest for 4,030 miles to Midway Island. The northernmost volcanoes have been worn down to form underwater mountains. The islands get older as you go north. The oldest ones are over 65 million years old. Only the volcanoes on the island of Hawaii are still active. In 1963, Canadian geologist Thomas Wilson suggested that the islands were formed over a "hot spot," where molten rock rose to the surface from deep inside the earth. The ocean floor has moved slowly over this hot spot. New volcanoes form as it moves, like the fire fountain on Hawaii (below), while old ones become extinct as they are moved away from it.

hot rocks rise toward the surface, then spread sideways and sink again as they cool. These movements of the mantle rocks are called convection currents. They carry the plates along with them. Where the mantle rocks are rising toward the surface, they will break through wherever there is a line of weakness.

The ever-spreading seabed

Running across the floors of the oceans are huge underwater mountain ranges called mid-oceanic ridges. These have been formed by underwater volcanoes and other splits in the ocean floor through which molten lava pours. In places the mountains rise so high that they form islands. Iceland is an example. Here, the volcanoes are still active, and new ones form from time to time.

If you measure the age of rocks on either side of these lines of volcanic activity, you find that the rocks get steadily older as you go away from the mountain ridge. New ocean floor is being formed along the ridges. As this happens, great stresses and strains break the rocks, forming a rather jagged pattern.

Splitting continents

Similar structures are found in a few places on land, too. The Great Rift valley, which extends roughly north to south from East Africa to the Red Sea, is also an area of spreading crust. East Africa, too, has its volcanoes and lava flows. The northeast corner of Africa is slowly splitting away from the rest of the continent. The mid-oceanic ridges and the rift valleys are the key to how the continents move.

Rising mountains, sinking seabed

If the plates along the mid-oceanic ridges and in the rift valleys are growing wider, where are they to go? Surely they will bump up against other plates?

Plates do indeed get forced underneath each other in places. The tremendous forces that occur where two plates collide in this way often push up sediments from the seabed to form great mountain ranges. The Himalayas, for example, are still forming today as the plate bearing India (and Australia) pushes against the Europe-Asia plate. The Andes are rising as the plates

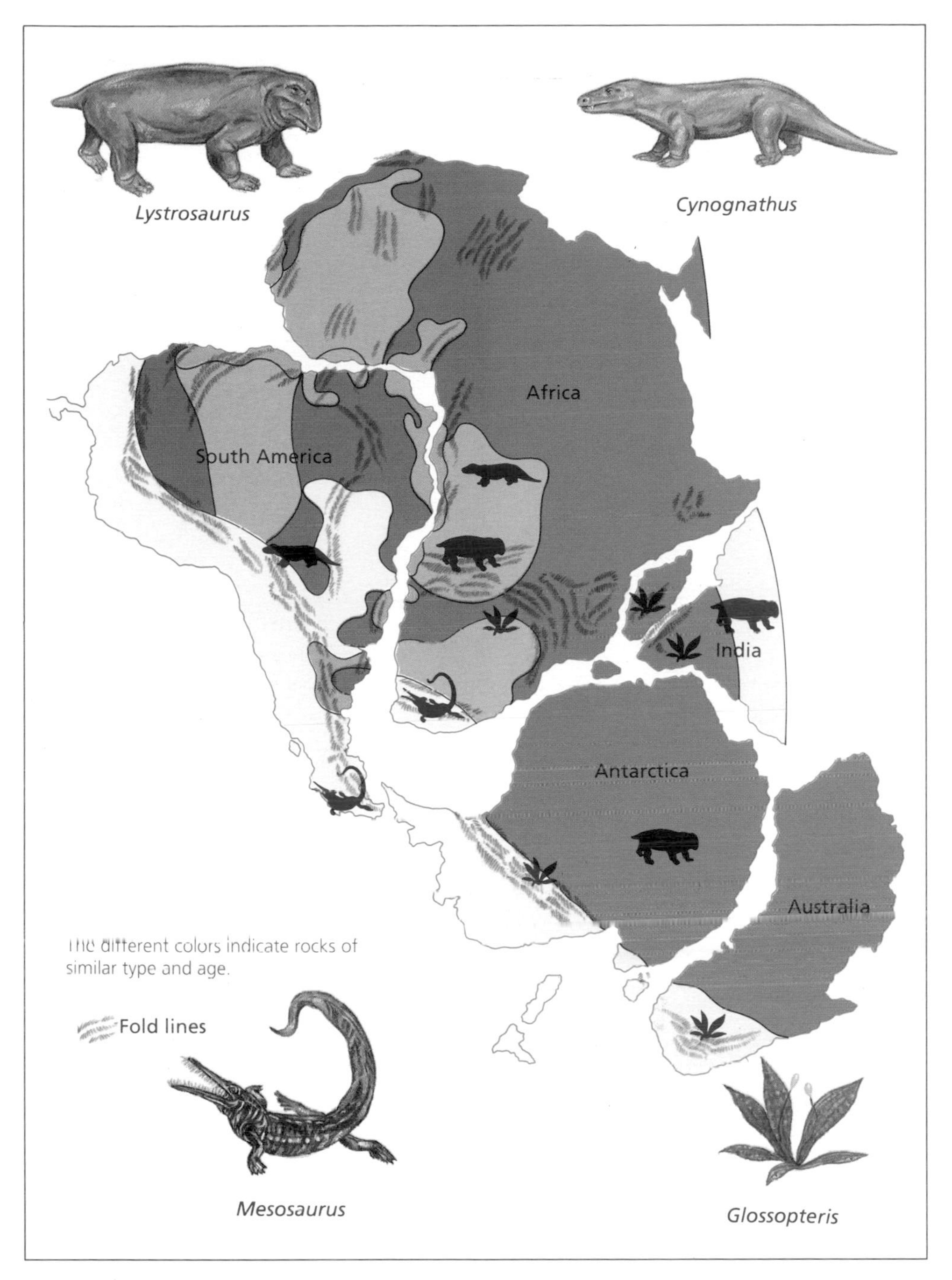

◀ Before the ancient supercontinent of Pangaea began to break up, the continents were all clustered together in the Southern Hemisphere. Evidence for this is to be found in the matching rock types and the fossils they contain. Since reptiles such as *Lystrosaurus*, *Cynognathus*, and the freshwater *Mesosaurus*, and plants like *Glossopteris* could not have crossed the oceans, the only explanation for their present distribution in the rocks is that these continents must once have been linked by land.

bearing the Pacific Ocean floor are forced beneath the western edge of the South American plate. The frequent earthquakes along the Californian coast are caused by rocks moving and fracturing as plates push past each other. The forces and processes that drive continental drift are called plate tectonics.

Vanishing rocks

New rocks are being formed and old rocks are disappearing all the time. As soon as mountain ranges begin to rise into the air, erosion begins. Rainwater, which is slightly acidic, dissolves certain minerals, weakening the rocks. Water in cracks and crannies freezes and expands, breaking up the rocks. In hot sun the surface layers of the rock expand and flake off. The loose rock fragments are swept away by rain, streams, and rivers, or become embedded in the ice of glaciers, scouring away more rock as they go. Eventually they are worn down to sand and mud, and carried to the sea by rivers or glaciers. They sink to the ocean floor, forming deep layers of sediments. Over millions of years these sediments become new rocks, which may one day be raised up to form new mountain ranges.

New rocks are also pouring out of the earth in the form of lava through volcanoes and other cracks in the earth's crust. New crust is forming where lava wells up along the mid-oceanic ridges. Where two plates meet, one is forced underneath the other, and its rocks are eventually absorbed into the mantle rocks below. These rocks, in turn, may rise to the surface as molten magma to form new crust. This constant renewing of the earth's crust is sometimes called the rock cycle.

Continents and climate

The changing distribution of land and sea has in turn affected the climate. Land heats up and cools down faster than water. The centers of very large land masses can become extremely hot or very, very cold. This affects the air pressure in the atmosphere above them, which in turn affects the world's weather.

In the world today, the North Pole lies under a frozen ocean, surrounded by land. Ocean currents bring warm water from the equator up the east coasts of North America and Asia toward the Arctic. But in the Southern Hemisphere the continent of Antarctica sits right on the South Pole, surrounded by ocean. There the ocean currents sweep round and round the continent, keeping it very cold.

The sea rises and falls

As the world climate changes and the polar ice expands and shrinks, the sea level falls and rises, too. When the sea level is low, land bridges form between the continents, like the link between North and South America. When the sea level rises, these links disappear and new islands are created. During the Recent Ice Age, the British Isles were joined to Europe. The Irish Sea and the English Channel formed as the ice melted.

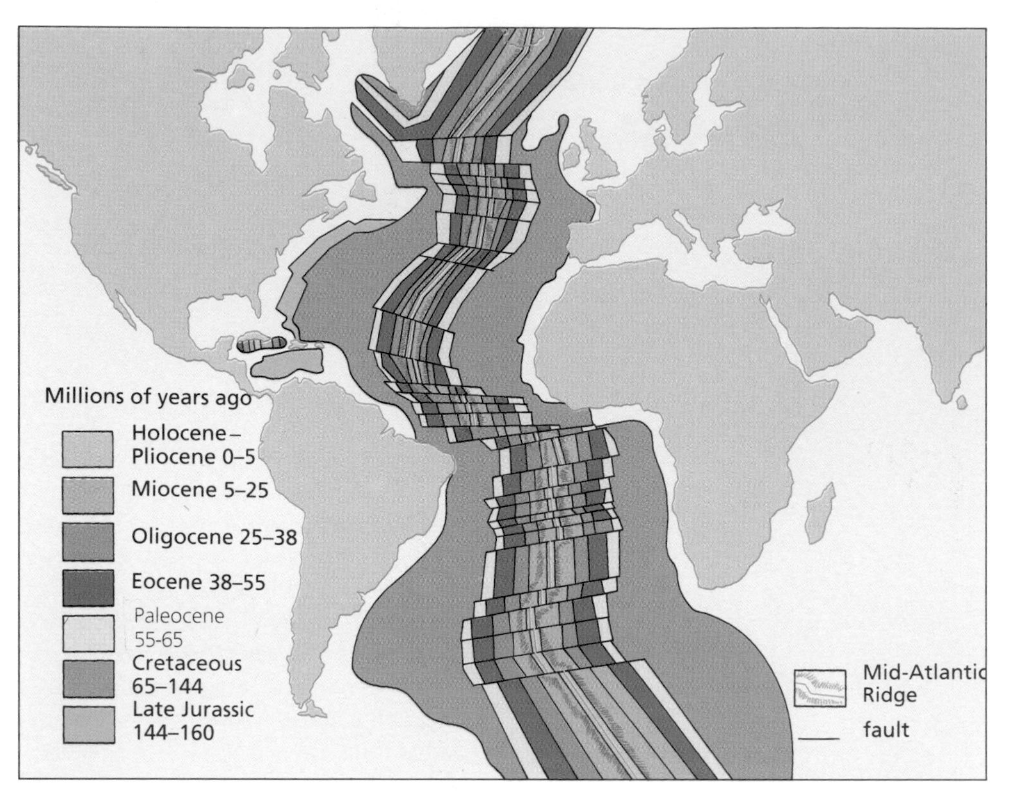

▲ New rock in the form of lava is welling up along the Mid-Atlantic Ridge and other mid-oceanic ridges, solidifying to form basalt. Evidence for this comes from dating the rocks of the ocean floor: The rocks are older the farther you go from the ridge. The newly forming rock pushes the older rocks away, causing great stresses, which result in numerous faults and fractures. Such forces help to move the plates around the globe, in the process that is known as plate tectonics.

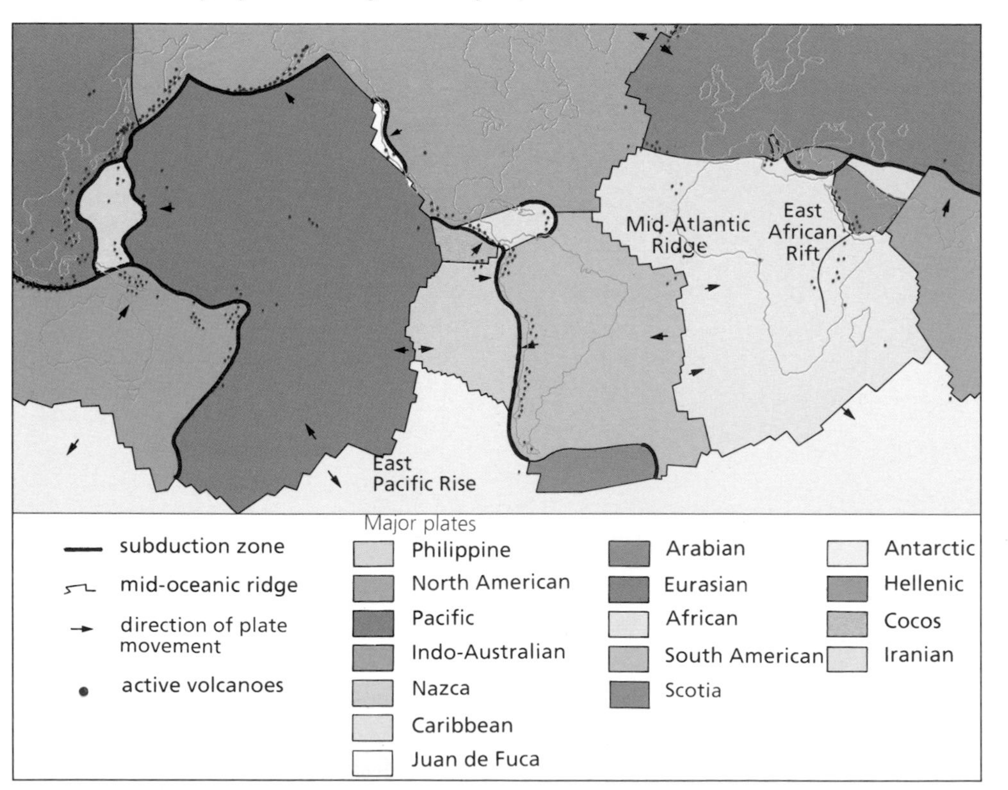

▲ The earth's crust is formed of a series of plates that are moving very slowly. Where two plates collide, they may simply buckle up at the edges, or one plate may be forced underneath the other. Running across some of the plates are rift valleys or mid-oceanic ridges. Here, new crust is added to the plates. This can lead to earthquakes and sometimes volcanoes, too.

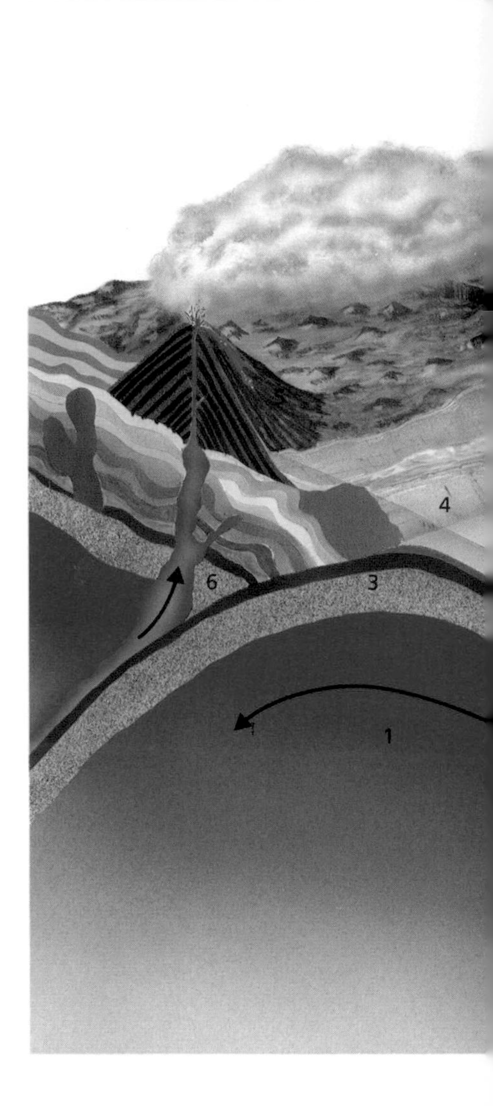

1 Convection currents in the mantle
The rocks of the earth's mantle are very hot, and they are under great pressure. This makes them become rather like very thick syrup: They can flow. As the hot rocks flow toward the surface, they cool and start to sink again. Where hot rocks are rising, they will well up into the crust wherever it is rather weak, to form mid-oceanic ridges, rift valleys, and chains of volcanic islands. These convection currents are the driving force behind the movements of plates.

2 Mid-oceanic ridges
Mid-oceanic ridges form a huge underwater mountain range over 4,030 miles long and up to 3,100 miles wide. In places they rise up to 11,480 feet from the seabed. Where they rise above the ocean surface, they form volcanic islands. Iceland is an example. Along the ridges hot molten rock wells up to form new crust. As more and more new rock forms along the ridge, the rocks that formed earlier

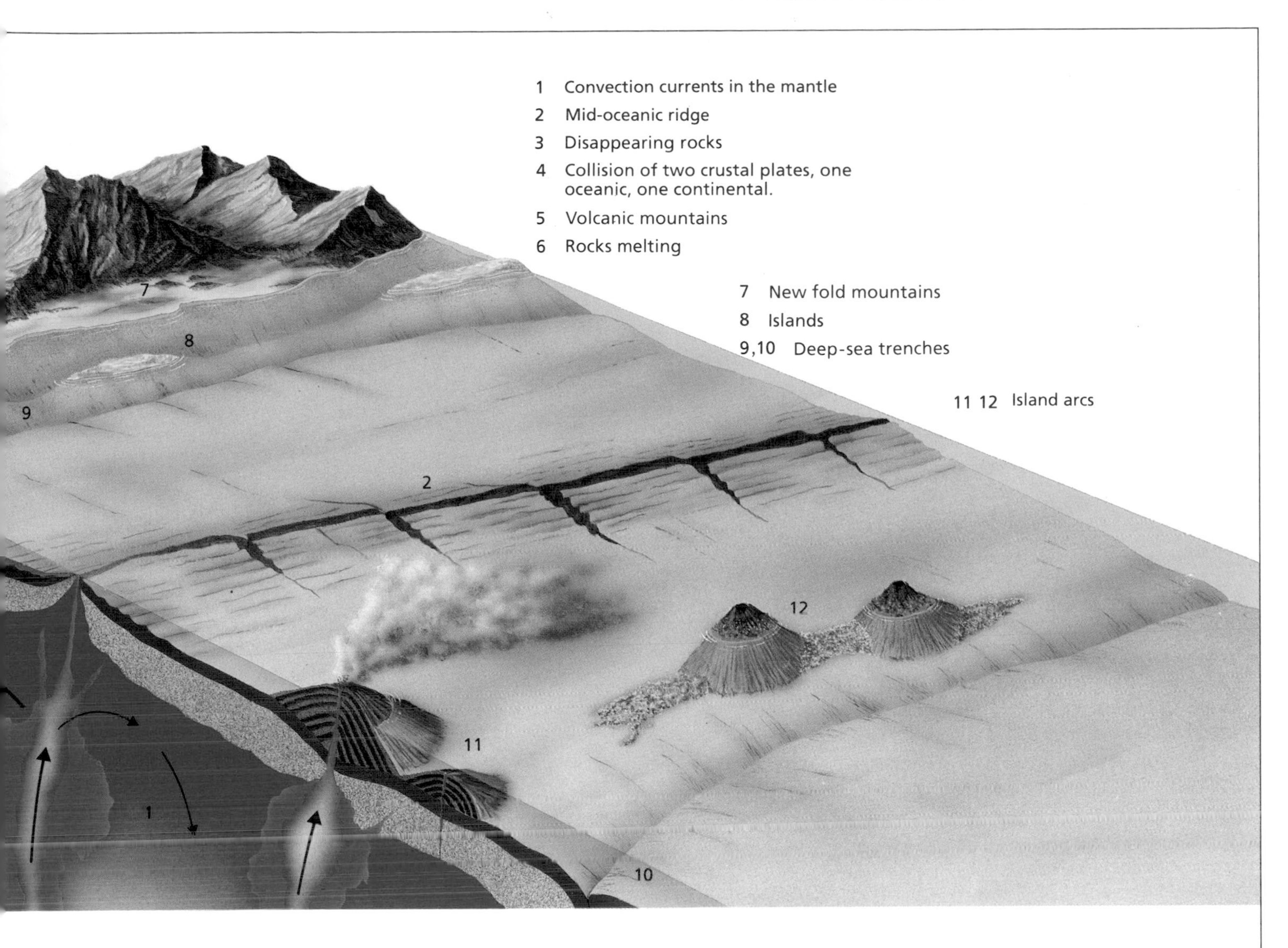

are pushed farther away. Gradually, the plate grows, pushing against neighboring plates. This process is called seabed spreading.

3 Disappearing rocks
As the new rocks formed at the mid-oceanic ridge are pushed farther away from the ridge, they become covered in sediments produced by the erosion of the ridge.

4 Collision of two plates
Here, one plate is being forced underneath the edge of the next plate. The tremendous forces that bring this about often cause a lot of earthquakes.

5, 6 Volcanic mountains
As the rocks of one plate are forced deep underneath the next (62 miles or more down), they become much hotter and are under great pressure. This makes them softer, until they are able to flow: they become magma. The magma rises toward the earth's surface along weak parts of the crust above to form volcanoes. There are many active volcanoes (and earthquakes) all around the North Pacific due to the pressures caused by the expanding Pacific plate.

7 New fold mountains
The great pressures generated where two plates collide are enough to fold and push up great thicknesses of sediments to form new mountain ranges.

8 Islands
Smaller folds offshore form chains of islands.

9, 10 Deep-sea trenches
Where a plate bearing part of the ocean floor is pushed under the plate next door, a deep trench forms. These ocean trenches may be five to seven miles deep and thousands of miles long. The Marianas Trench in the Pacific Ocean is more than seven miles deep.

11, 12 Island arcs
Magma may well up to the surface from "hot spots" in the mantle, forming volcanic islands. As the plates drift across the hot spot, old volcanoes become extinct and new ones form, creating a chain of volcanic islands. Some of these may eventually be worn down, becoming underwater mountains called seamounts.

Fossils, Nature's Own Clues

▲ Some of the best-preserved fossils are insects and other small creatures trapped in amber. These flies are more than a million years old, yet much of their internal structure has been preserved intact.

The ancient Greek philosophers puzzled over fossils. They found seashells set in stone high up in the mountains, and guessed that they had once been living animals. This means, said the philosophers, that this area was once covered by the sea. Correct! But how did the fossils come to be there? How did the shells get into the rocks?

Fossils are the remains and traces of plants and animals that lived long ago. But very few plants and animals ever turn into fossils. Their remains may be eaten or attacked by fungi and bacteria, and soon there is nothing left. If they have a shell or a bony skeleton, this may survive longer, but eventually it, too, will be broken up. Only if the remains are buried very rapidly, before they have had time to decay, will they have a chance of surviving as fossils.

Turned to stone

When a dead plant or animal is buried rapidly, sediments like sand or mud are piled on top. The remains are soon buried out of reach of the air, so they will not rot away. Over millions of years, the pressure of the sediments above turns the lower ones into rock. Water seeping through the sediments contains minerals, which have sometimes been dissolved from the sediments themselves.

Eventually this water is squeezed out by the weight of the sediments above. But the minerals remain to bind the sediments together and help them harden into rock. These minerals are also deposited in the plant and animal remains, filling the spaces between their cells, and sometimes even taking the place of the bones and shells. The creatures become preserved in stone. Millions of years later, colliding continents may force these rocks up out of the sea to become dry land. Rain, wind, or perhaps the sea will wear them away until the fossils are exposed.

Perfect fossils

Some of the best-preserved fossils are insects and other small creatures trapped in amber. Amber is a sticky resin that oozes out of the trunks of certain kinds of trees when they are damaged. Its aromatic smell attracts insects, which get stuck to the resin and become imprisoned in it.

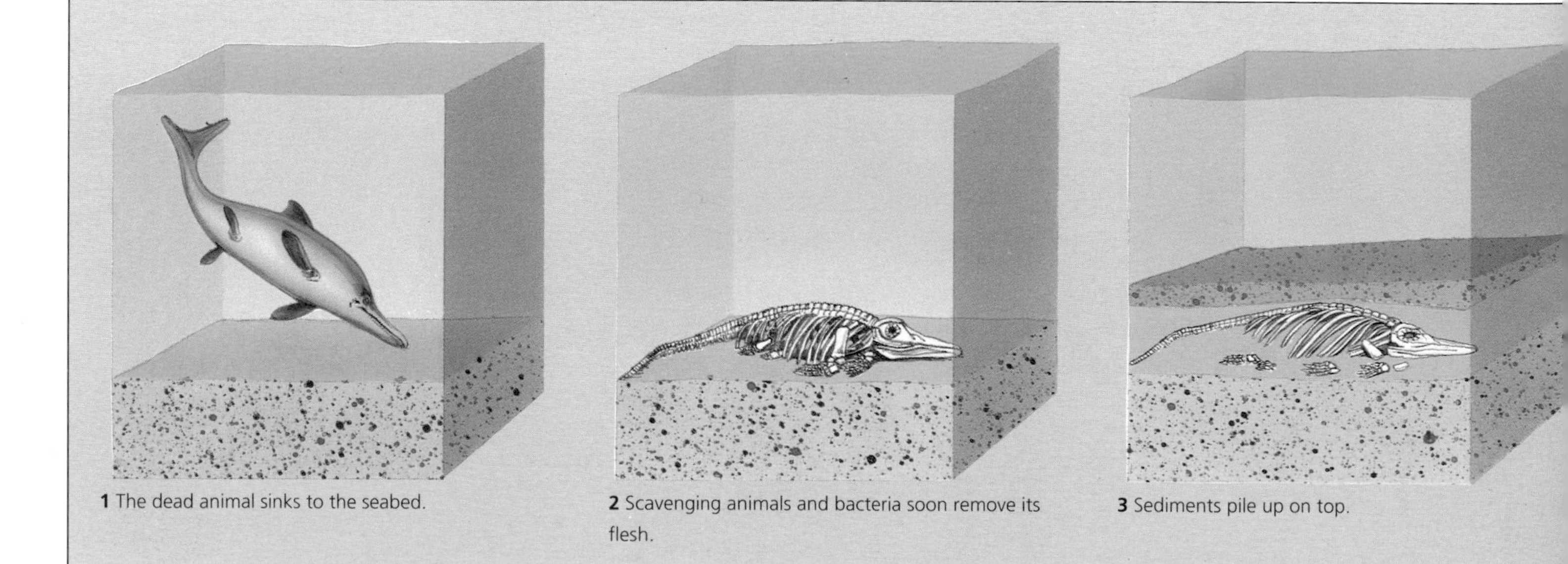

1 The dead animal sinks to the seabed.

2 Scavenging animals and bacteria soon remove its flesh.

3 Sediments pile up on top.

▶ A dinosaur footprint preserved for millions of years in rocks at Moenawe, Arizona.

The resin sets to form a hard, transparent substance that protects the remains inside from decay. The delicate structures of ancient insects and spiders have been found perfectly preserved in amber. It is even possible to extract their genetic material (DNA) and analyze it.

Some of the most delicate and beautiful fossils are found in rocks associated with coal deposits. Coal is a hard black rock made mainly of carbon from the remains of ancient plants. It was formed in swamp forests millions of years ago. The swamp forests were sometimes overrun by the sea and buried rapidly under layers of mud, which was later hardened and compressed to form mudstones and shales.

The leaves and stems of the coal-forest plants are sometimes preserved as layers of coal, or as thin black films of carbon lying between the layers of shale. Sometimes just the imprints of tree bark, leaves, or fern fronds may remain in the rocks. Shales split easily along flat planes, and you can easily find the fossils of whole fronds lying on the newly exposed surfaces.

Even more remarkable are the fossils found in coal balls. These form where lime-rich water has trickled through the plant remains, encasing them in limestone as it evaporated, and preserving every minute detail of delicate plant structure.

Traces of the past

Even when an animal's actual remains are not preserved, it may leave its footprints behind. Sometimes the footprints are literally preserved in the sediments, for example footprints in sand that get filled up with mud and so preserved. Animals can leave other traces, such as grooves in the sediment made as they shuffle through the mud, graze on detritus (food fragments), or burrow into the bed of a lake or sea. These "trace fossils" tell us not only that the animals were there, but also something about how they lived and moved.

Hard-shelled animals such as trilobites and horseshoe crabs may leave a variety of trails in soft muds, depending on whether they are resting, walking, or feeding. Many of these trails have been given their own

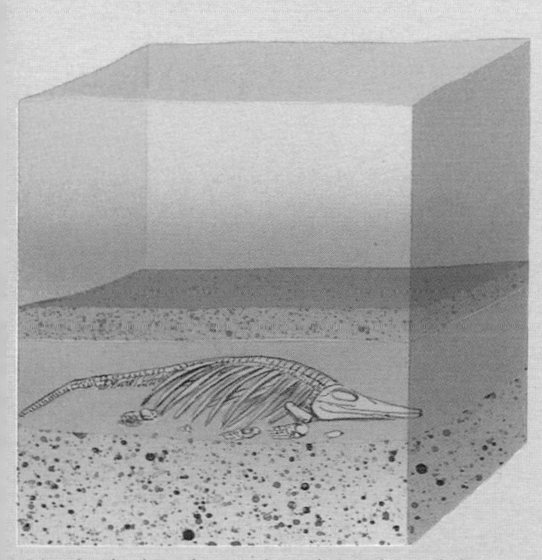

4 Dissolved minerals seep into the rocks and the remains.

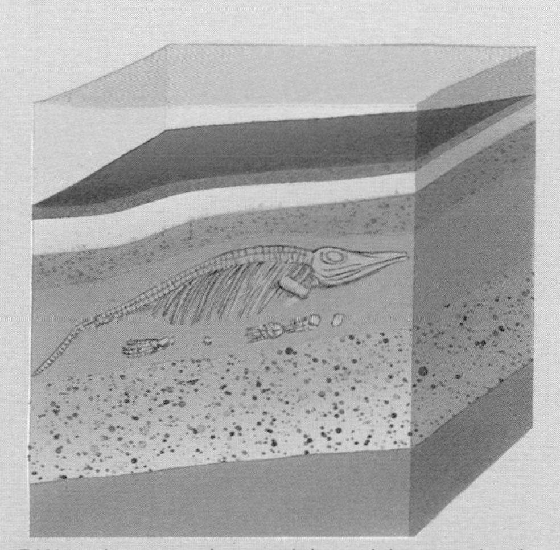

5 Water is squeezed out and the rock becomes hard and compact. The minerals from the water replace the chemicals in the bones.

6 Millions of years later the rocks are lifted up and become dry land. Rain, wind, or perhaps the sea wear them away until the fossil is exposed.

CAST AND MOLD FOSSILS

Sometimes the water in the sediments dissolves away the remains of a buried organism, leaving a hole of exactly the same shape. This is a mold fossil (left). The mold may then fill with minerals, forming a cast fossil that has the same shape but no internal structure (right).

special names by scientists who did not know which animals had made them.

Sometimes the animal's droppings are fossilized. They may even be preserved well enough for a scientist to find out what the animal had been eating. In well-preserved fossils, food may still be in the animal's stomach. Ichthyosaurs, dolphin-like marine reptiles, have been found with whole fish inside them – the remains of recently swallowed meals they did not have time to digest before they died.

Footprints in stone

Fossilized dinosaur footprints have told us a lot about how dinosaurs moved and lived. Fossil tracks show how far apart their feet were. This tells us whether their legs were splayed out to the sides of their bodies like those of modern lizards, or whether they were vertical, giving more support to the body. It is even possible to calculate how fast the dinosaur walked.

Scientists can also tell which dinosaurs trailed their tails as they walked and which ones held them up. In parts of the United States there are fossil pathways of various carnivorous (meat-eating) and herbivorous (plant-eating) dinosaurs that have lots of footprints all traveling in the same direction. These dinosaurs traveled in herds. The size of the footprints shows how many young there were and where in the herds they traveled.

Getting the picture

The study of fossils is called paleontology, which in Greek means "the study of ancient life." Unfortunately, using fossils as clues to the past is not so easy as the pictures in this chapter might suggest. Even in the rare moments when plant and animal remains are buried fast enough to be preserved, they seldom remain undisturbed. Water currents may sweep them into heaps, breaking them apart. The heavier parts sink into a different position, and the lighter parts are washed away. Floods and landslides may stir up the sediments. Some plants and animals stand almost no chance of being preserved because they live in habitats where little sediment is laid down. The inhabitants of forests or grasslands, for example, are extremely unlikely to be swept away by floods or buried by sand or mud to become fossils.

Just as detectives need to know whether or not a body has been moved, so paleontologists need to know whether the fossils found in a particular place were from animals that actually died in that place and position. If they did, they are called a life assemblage. Life assemblages can tell us which animals lived together. From this it is often possible to guess what their habitat was like – whether it was underwater or on land, warm or cold, wet or dry. The rocks, too, provide more clues about the ancient environment. But all too often the remains are washed far from their first resting place, and are broken up on the way. Land animals may even be swept out to sea, confusing the detectives. These collections of fossils, which reached their final resting place far away from where the original animals and plants died, are called death assemblages.

What did the hedgehog look like before it crossed the road?

Part of the fun of paleontology is trying to piece together a fossil from the few bits that have survived. When the ancient animal was unlike any animals alive today, this can be difficult. Often in the past, different parts of the same animal have been given different names, and were thought to be different kinds of animals.

Fossil hunter's delight – lots of ammonites and bivalve shells in one place. This is a death assemblage: The fossils were not preserved where the animals once lived, but were swept away by water currents and deposited in a heap somewhere else, where they were rapidly buried. These animals lived some 150 million years ago during the Jurassic period.

The scientists who first studied fossils from the ancient Burgess shale in the Canadian Rockies, which is 570 million years old, found several puzzling specimens. One appeared to be the rather odd tail end of a shrimp. They called it *Anomalocaris*, which means "odd shrimp." Another fossil was like a flattened jellyfish with a hole in the middle, which they called *Peytoia*. A third fossil, which they called *Laggania*, looked like the squashed body of a sea cucumber. Later they found fossils of *Laggania* and *Peytoia* together, and decided they were really a sponge with a jellyfish sitting on top.

These fossils were put away in museum drawers and forgotten until a few years ago. Then a new generation of paleontologists took them out and began to study them again. They realized that all three kinds of fossils were often found together in the rocks. Could they be related? They looked at as many of these fossils as possible and came to a startling conclusion. They were all different parts of the same animal, a very "odd shrimp" indeed, and probably the largest animal in the sea at that time. It was like a huge legless shrimp, up to 25 inches long, with an oval head (*Tuzoia*), two large eyes on stalks, and a large round mouth (*Peytoia*) surrounded by hard teeth. It had a pair of feeding limbs (*Anomalocaris*) in front up to 7 inches long. *Laggania* appeared to be the flattened remains of the rest of the body.

Bringing fossils to life

If you know how to read the clues in the rocks, you can discover many fascinating facts about how creatures lived in the past. Ammonite shells with what may well be mosasaur (a large marine reptile) tooth marks on them tell of past attacks by other animals. The tooth marks of rodents on fossil mammal bones show that the rodents scavenged on their corpses. Starfish have been found fossilized while feeding on beds of mollusks, and lungfish have been preserved while lying dormant in their burrows in the mud. Baby dinosaurs have been found in the act of hatching from eggs. But these are all rare finds. Usually we have to use our knowledge of how the modern descendants of these animal groups live to deduce how their ancestors lived.

Hunting for fossils

It is surprising in how many different places you can find fossils today – not only in cliffs and quarries, but in the stone of city walls, in building rubble, and even in the

1

2

3

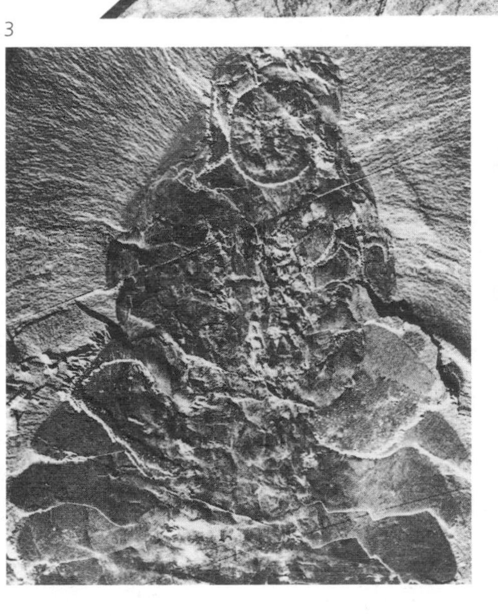

4

▶ The history of the fossil *Anomalocaris* shows how difficult it can be to reconstruct fossils from the fragments that have survived. *Anomalocaris* (1) was a large, strange, shrimplike creature that lived in the early Cambrian period. For many years the only remains known were fragments so distinctive that at first they were thought to be quite different species; the original *Anomalocaris* (2) turned out to be a mouthpart, *Laggania* (3) the body, and *Peytoia* (4) the mouth.

◀ The petrified remains of a Triassic forest in Petrified Forest National Park, Arizona. Forests can be turned to stone if they are suddenly covered by the sea. Minerals from the seawater penetrate the wood and crystallize to form rock. You can sometimes see the mineral crystals in the tree trunks, turning them beautiful shades of red and purple.

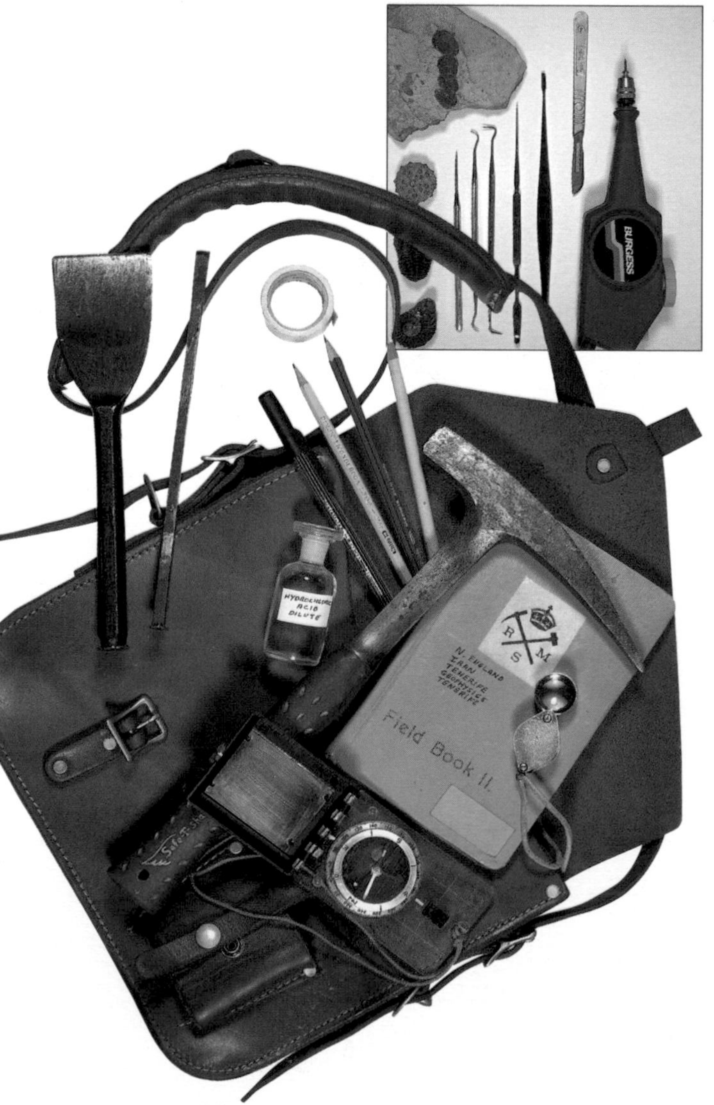

▶ Tools for fossil hunting. The head of a geological hammer has a special flat face for hitting rock faces, and a wedge-shaped end for levering rocks apart. You can also use stone chisels, which come in various sizes. A notebook and compass are essential for recording the exact position of fossils in the rocks, and the orientation of the rocks in the quarry or cliff. A hand lens will reveal tiny fossils such as fish teeth or scales. Some geologists carry diluted acid to help extract delicate fossils, but this is best done in the laboratory. Here, more delicate operations require a variety of fine picks, needles, and scrapers. The electric tool is a vibrator, which can help to loosen the rock.

stones in your garden. But they are found only in sedimentary rocks: limestone, chalk, sandstone, mudstone, slate, or shale.

The best way to become a good fossil hunter is to learn from the experts. Find out from your local library if there are any geological or natural-history clubs that arrange outings to look for fossils. They will take you to the best places and show you where the fossils are.

Do your homework

Like a detective, you need to find out more about the clues you are hunting for. Visit the local library and find out what rocks there are in your area. The library will have maps showing the rocks. How old are these rocks? What fossils would you expect to find in them? Visit your local museum and see what fossils other people have found in your area. Often you find only parts of fossils, and they are easier to spot if you know what you are looking for.

Playing safe

It is important to be properly prepared for fossil hunting. Poking around at the foot of a cliff or in a quarry can be dangerous. First you must write to the owners to get permission to visit. They will be able to warn you of any special dangers. Quarries and cliffs are dangerous, lonely places, and you should not visit them alone. Always leave a message with someone telling him or her where you have gone.

You should wear a hard hat – a bicycle helmet will do. When you hit the rocks with a hammer, you should also wear protective goggles or glasses because tiny flakes of rock may fly off at great speed and they could blind you. Do not try to hammer fossils out of a cliff. The vibrations can easily loosen the rock above you, and start a rockfall. You will usually find plenty of fossils in the rocks already on the ground.

Your own geological record

A good amateur geologist always makes notes. It is important to know exactly when and where you found a fossil. This means not just the name of the cliffs, quarry, or building site, but where you found the fossil. Was it in a large piece of rock or a small piece; near a cliff or in the ground itself? Were there any other fossils nearby? If so, what were they? In which direction were the fossils pointing? This may tell you more about how the animal lived and died. Try to make a drawing of the place where you found it. Graph paper will make this easier. You may like to take a photograph, too, but drawings often show detail better.

Photographs and drawings can also provide souvenirs of fossils you cannot take home. You may be able to make a plaster cast of a fossil, or use Plasticine to make a mold. Even if the fossil remains firmly fixed in the rock, it can still tell you something about the history of the place.

Remember to take containers with you for carrying the fossils. You can wrap large fossils in newspapers or plastic bags. Put small fossils in margarine tubs filled with cotton. Take some labels for the boxes and for the fossils. You will be surprised at how quickly you forget where you found each fossil.

Tools of the trade

To break open the rocks and get the fossils out, you need a geological hammer (one with a large flat end). A set of chisels made for use on stone will help you remove any unwanted rock from your fossil. Go carefully – it is all too easy to smash the fossil. An old knife will scrape away soft rock, and a toothbrush can clean away dust and small rock particles.

Professional fossil hunters, or paleontologists, take the rocks containing the fossils back to the laboratory. If the fossils are delicate or crumbly, they may protect them in layers of plaster of Paris or plastic foam before they cut away the rock. Back in the laboratory, the scientists may use dentists' drills, high-pressure water jets, and even acids to extract their fossils from the rock. Often they protect the fossil with chemicals to make it harder before they start to work around it. At every stage, they make careful measured drawings and take photographs of both the fossil and its surroundings.

CLUES FROM FOSSILS

Environment Fossils provide clues to the kind of environment in which the rock was formed.

Climate Fossils can tell us what the ancient climate was like.

Evolution Fossils tell us how living things have changed through the ages.

Dating the rocks Fossils help us to tell the age of rocks, and to track the movements of the continents.

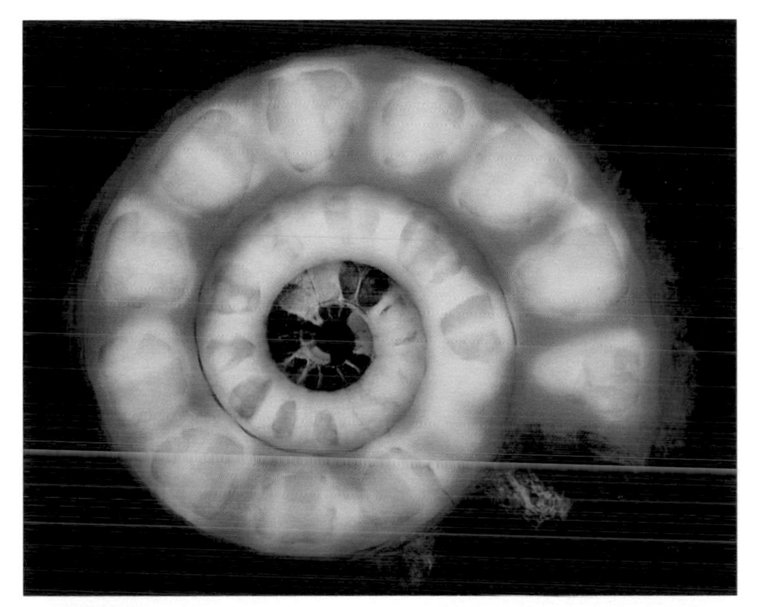

◀ An artificially colored X-ray photograph reveals the internal structure of a fossil ammonite. You can see the walls separating the internal chambers.

▼ A geologist uses a very fine chisel to extract fossil dinosaur bones from rocks at Dinosaur National Monument in Utah.

The Story of Claws

In 1983 an English amateur fossil hunter, William Walker, was looking for fossils in a Surrey clay pit when he noticed a large ball-shaped rock with a small piece of bone sticking out from it. He broke it open with his hammer, and out fell the pieces of a giant claw almost 14 inches long. He sent it to the Natural History Museum in London, where they realized it was a very exciting find – the claw of a flesh-eating dinosaur. The museum sent a team of scientists to explore the clay pit, and they managed to dig up many more bones – over 2 tons in all. They nicknamed the dinosaur Claws.

Saving Claws

To protect the bones from drying out and cracking, they wrapped some of them in plaster of Paris bandages. Others were wrapped in aluminium foil and covered in plastic foam. Special machines were used to remove the rock from around the fossil. Then the bones were made tougher by soaking them in resin. Copies were made in fiberglass and resin to send to other museums.

Putting Humpty-Dumpty together again

When they fitted the bones together, the scientists found that they had discovered a new kind of dinosaur, which they called *Baryonyx walkeri*. *Baryonyx* is Greek for "heavy claw," and *walkeri* is in honor of its discoverer, William Walker. *Baryonyx* was 30-33 feet long. It probably stood on its hind legs and was about 13 feet tall. Claws must have weighed about 2 tons each. Its long narrow snout is armed with lots of teeth, rather like the snout of a modern crocodile, which suggests that it was a fish eater. Fish teeth and scales were found in the dinosaur's stomach. The long claw was probably on its thumb. We do not know if it used the claw to catch the fish, or whether it caught them in its mouth like a crocodile. The clay pit where Claws died 124 million years ago was at that time a lake in a large river floodplain surrounded by marshes full of horsetails and ferns. When Claws died, its corpse was washed into the lake, where it quickly became covered in fine mud and silt. Several kinds of plant-eating dinosaurs have been found in these rocks, including the large *Iguanodon*. But *Baryonyx* is the only flesh-eating dinosaur known from rocks of this age anywhere in the world. Similar bones were found 30 years ago in the Sahara Desert, so *Baryonyx*-like dinosaurs probably ranged over a large area stretching from England to North Africa.

▲ Paleontologists use plaster of Paris to protect bones and prevent them from crumbling as they are transported back to the museum. Bandages soaked in wet plaster are wrapped around the fossil or the fossil-bearing rocks.

▼ A paleontologist uses a diamond-edged dental saw to cut away the rock around a dinosaur vertebra. He will then use a finer engraving tool to remove the last traces of rock.

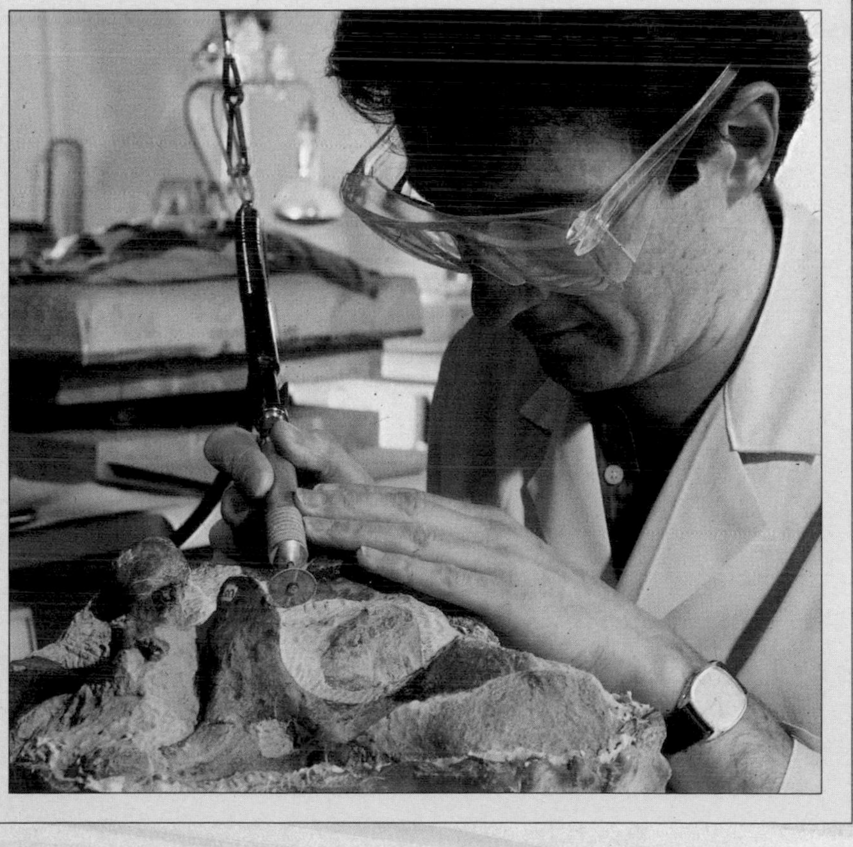

LIVING FOSSILS

What do a ginkgo, a coelacanth, a horseshoe crab and a nautilus have in common? All belong to groups of organisms that have been around for millions of years. All have changed very little over long periods of geological time, and all have features that seem "primitive" when they are compared with most modern groups of plants and animals. And all have very few living relatives. They are all living fossils.

On December 23, 1938, a young South African museum curator, Marjorie Courtenay-Latimer, was summoned to the beach to see a strange and very bad-tempered fish that had been caught by local fishermen. It was a large fish, 5 feet long, but the first thing Marjorie noticed was its color, a pale purplish blue with silver markings. She had never seen anything like it before.

But how to get it back to the museum? It was almost Christmas, and the local taxi driver was reluctant to have a "stinking fish" in his taxi. Threats to call another taxi won him over, but it was still a struggle to carry the fish even a short distance because it weighed 128 pounds. Christmas is also the South African summer, and in those days refrigerators were rare. The fish began to rot at a fast rate. Marjorie sent an urgent letter with a sketch of the fish to fish expert Professor James Leonard Brierly Smith, 250 miles away in Grahamstown, but it was January 3, 1939, before he received it.

Brierly Smith stared at the sketch. He had seen something like this somewhere before . . . but where? Suddenly it dawned on him: He was looking at something from the distant past, something he had seen only in sketches of ancient fossils – a creature that had been believed extinct for almost 100 million years. A coelacanth! His hunch was confirmed in February, when he finally got to see the fish. The news was telegraphed throughout the world: "MISSING LINK FOUND!"

WANTED!

Where there was one coelacanth, there must be more. The search was on for more information and more specimens. A reward was offered; posters and leaflets were sent out all over South and East Africa. But no more coelacanths appeared.

Smith was puzzled. If coelacanths lived off the South African coast, the fishermen should be able to catch more. Perhaps the coelacanth had been swimming off course. Perhaps this was not its usual home. He studied the ocean currents, and found that there were strong currents moving southward from East Africa. Perhaps the coelacanths lived farther north.

A group of islands between Madagascar and mainland Africa, the Comoro Islands, caught Smith's attention. He decided to pass out some more leaflets

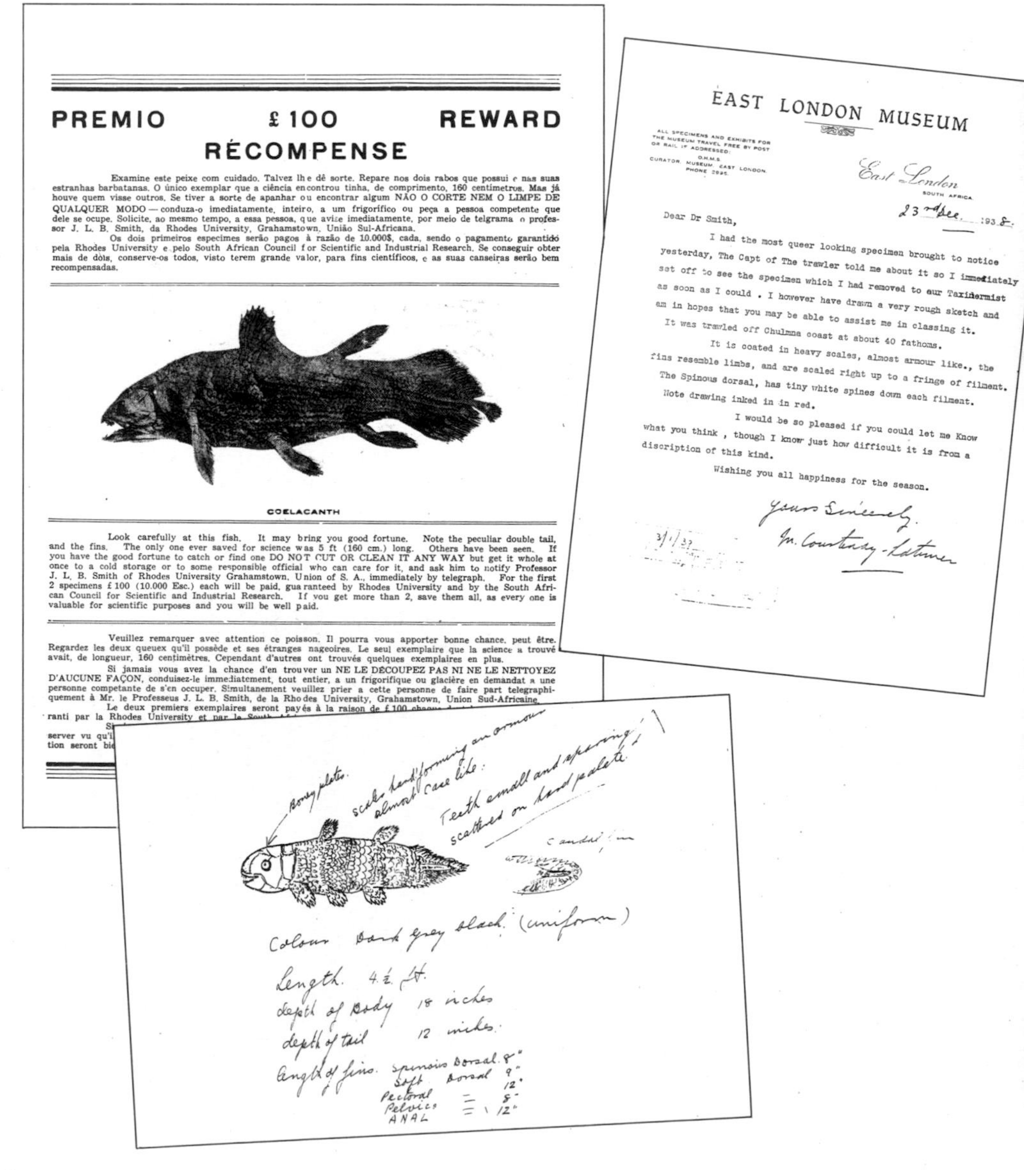

PREMIO £100 REWARD

RÉCOMPENSE

Examine este peixe com cuidado. Talvez lhe dê sorte. Repare nos dois rabos que possui e nas suas estranhas barbatanas. O único exemplar que a ciência encontrou tinha, de comprimento, 160 centímetros. Mas já houve quem visse outros. Se tiver a sorte de apanhar ou encontrar algum NÃO O CORTE NEM O LIMPE DE QUALQUER MODO — conduza-o imediatamente, inteiro, a um frigorífico ou peça a pessoa competente que dele se ocupe. Solicite, ao mesmo tempo, a essa pessoa, que avise imediatamente, por meio de telgrama o professor J. L. B. Smith, da Rhodes University, Grahamstown, União Sul-Africana.

Os dois primeiros especimes serão pagos à razão de 10.000$, cada, sendo o pagamento garantido pela Rhodes University e pelo South African Council for Scientific and Industrial Research. Se conseguir obter mais de dois, conserve-os todos, visto terem grande valor, para fins científicos, e as suas canseiras serão bem recompensadas.

COELACANTH

Look carefully at this fish. It may bring you good fortune. Note the peculiar double tail, and the fins. The only one ever saved for science was 5 ft (160 cm.) long. Others have been seen. If you have the good fortune to catch or find one DO NOT CUT OR CLEAN IT ANY WAY but get it whole at once to a cold storage or to some responsible official who can care for it, and ask him to notify Professor J. L. B. Smith of Rhodes University Grahamstown, Union of S. A., immediately by telegraph. For the first 2 specimens £100 (10.000 Esc.) each will be paid, guaranteed by Rhodes University and by the South African Council for Scientific and Industrial Research. If you get more than 2, save them all, as every one is valuable for scientific purposes and you will be well paid.

Veuillez remarquer avec attention ce poisson. Il pourra vous apporter bonne chance, peut être. Regardez les deux queuex qu'il possède et ses étranges nageoires. Le seul exemplaire que la science a trouvé avait, de longueur, 160 centimètres. Cependant d'autres ont trouvés quelques exemplaires en plus.

Si jamais vous avez la chance d'en trouver un NE LE DÉCOUPEZ PAS NI NE LE NETTOYEZ D'AUCUNE FAÇON, conduisez-le immediatement, tout entier, a un frigorifique ou glacière en demandat a une personne competante de s'en occuper. Simultanement veuillez prier a cette personne de faire part telegraphiquement à Mr. le Professeus J. L. B. Smith, de la Rhodes University, Grahamstown, Union Sud-Africaine.

Le deux premiers exemplaires seront payés à la raison de £100 [illegible]
ranti par la Rhodes University et par le [illegible]
S[illegible]
server vu qu'i[illegible]
tion seront bie[illegible]

EAST LONDON MUSEUM

ALL SPECIMENS AND EXHIBITS FOR THE MUSEUM TRAVEL FREE BY POST OR RAIL IF ADDRESSED:
O.H.M.S.
CURATOR, MUSEUM, EAST LONDON.
PHONE 2995.

East London
SOUTH AFRICA
23rd Dec. 1938.

Dear Dr Smith,

I had the most queer looking specimen brought to notice yesterday, The Capt of The trawler told me about it so I immediately set off to see the specimen which I had removed to our Taxidermist as soon as I could . I however have drawn a very rough sketch and am in hopes that you may be able to assist me in classing it.

It was trawled off Chulmna coast at about 40 fathoms.

It is coated in heavy scales, almost armour like., the fins resemble limbs, and are scaled right up to a fringe of filment. The Spinous dorsal, has tiny white spines down each filment. Note drawing inked in in red.

I would be so pleased if you could let me Know what you think , though I know just how difficult it is from a discription of this kind.

Wishing you all happiness for the season.

Yours Sincerely.
M. Courtenay-Latimer

3/1/39

Boney plates.
scales hard forming an armour almost case like.
Teeth small and sparing scattered on hard palate.
Colour Dark grey black. (uniform)
Length. 4½ ft.
depth of Body 18 inches
depth of tail 12 inches.
Length of fins. Spinous Dorsal 8"
Soft Dorsal 9"
12"
Pectoral = 8"
Pelvics = 12"
ANAL

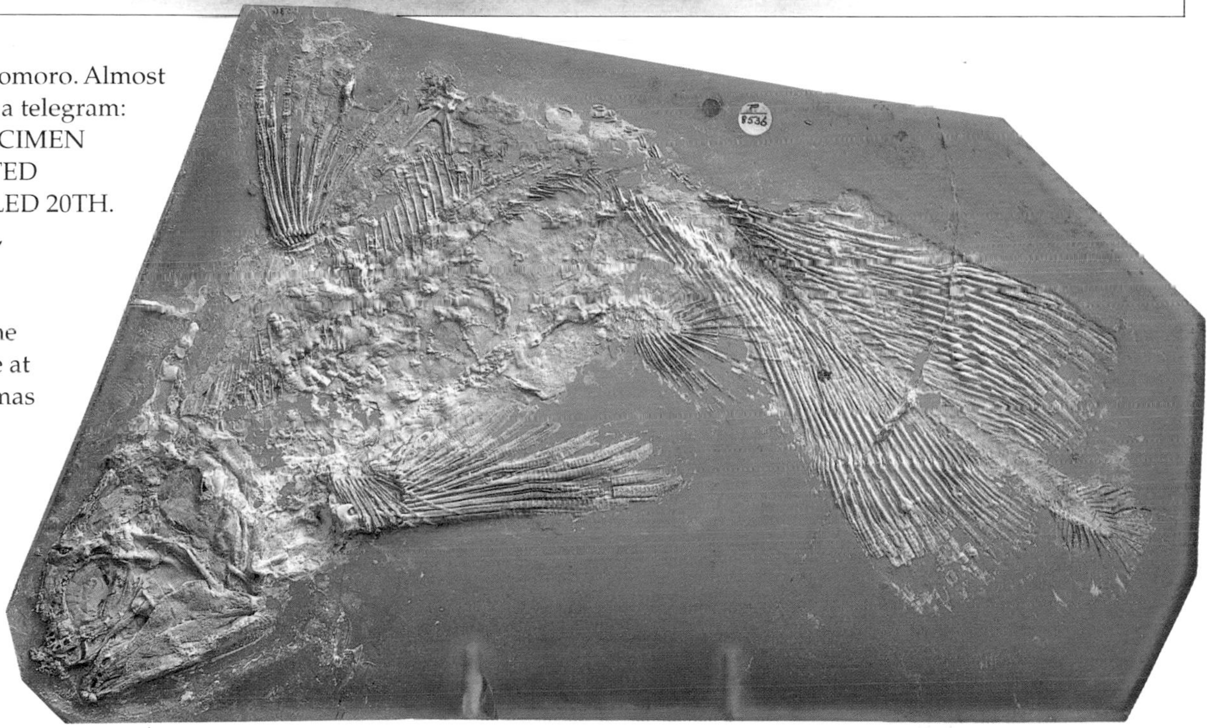

offering a reward in the Comoro. Almost immediately he received a telegram: "HAVE FIVE-FOOT SPECIMEN COELACANTH. INJECTED FORMALIN HERE. KILLED 20TH. ADVISE REPLY, HUNT, DZAOUDI."

Like the first coelacanth, the second one had put in its appearance at Christmas. It was Christmas Eve, 14 years since the first fish had been discovered.

Brierly Smith was thousands of miles away. In desperation he appealed to the prime minister of the Union of South Africa, Daniel Malan, who agreed to lend him a government plane to collect it.

Fossil gold

Soon more coelacanths were caught. They were now in great demand with the local fishermen. Museums were offering money, and before long they were also being sold as curios. There were even claims that they could be made into love potions.

The scientists discovered that the coelacanths live in deep water, between 600 and 2,000 feet. They are found in an area where fresh water in the rocks seeps out through underwater caves into the ocean – very special conditions indeed. This means that they may live in a very small area, so the population may not be very large.

Ironically the very discovery of the coelacanths may have doomed them. Coelacanths reproduce very slowly. They produce huge eggs – as big as grapefruits – that are kept inside the female until they hatch. This means that not many eggs can be produced at a time. So, even though the young are born as miniature coelacanths with a very good chance of survival, this slow rate of reproduction means that coelacanths could easily be overfished.

▲ Reconstruction of an ancient coelacanth (top). Coelacanths have changed very little for millions of years. From the fossil (above), you can see the heavy head, the bony supports of the fleshy fins, and the long central part of the tail, all features of living coelacanths.

◀ A living coelacanth swimming. Notice how one front leg is directed forward, the other backward. Coelacanths use their fleshy fins in much the same way as four-legged animals use their legs, moving them forward and back in the same pattern, but in this case they are acting as paddles. It is thought that four-footed vertebrates – amphibians, reptiles, and mammals – are descended from ancestors of the coelacanths.

Old four-legs

The coelacanth belongs to the ancient order of lobe-finned fish, the Sarcopterygii. The pairs of pectoral and pelvic fins (the fins behind its eyes and on its belly) are at the tips of leglike stumps supported by small bones, and so are the second dorsal (on its back) and the anal (tail) fins. The tail fin is in three parts; the middle part is on a short stalk.

The really special thing about the coelacanth is its fins. Scientists have been able to film the coelacanth swimming and feeding in its natural habitat. It uses its paired fins in much the same way as modern newts, lizards, and dogs use their legs for walking: Diagonally opposite legs move forward one after the other, then the other pair of legs move forward. Instead of walking on the ground, the legs serve as paddles as it feeds on fish and squid. Sometimes it even swims backward or head down.

Missing link or blind alley?

Nobody is quite sure where the coelacanth fits into the story of evolution. Some paleontologists think it may be related to the ancestors of the amphibians, a "missing link" between the fish and the amphibians. Others think it is a blind alley of evolution, and belongs to an ancient line that all but died out long ago.

In the Devonian period, 400 million years ago, coelacanths were widespread. They lived in freshwater lakes and in the open ocean. There are still many unsolved mysteries surrounding the coelacanth. Why did almost all of them die out? And why did a few survive just off the Comoro Islands? What was special about that place? It would be a pity if, after surviving for 400 million years, the coelacanths now become extinct because of the whims of tourists or the demands of museums.

Plants from the past

The world's largest living thing – the giant redwood – is a survivor from the days of the dinosaurs. Herds of long-necked sauropod dinosaurs probably browsed forests of redwoods, whose descendants are now the tallest trees in the world. The dawn redwood tree was known only from fossils until 1948, when living specimens were discovered in central China.

The ginkgo has an even older history. Similar trees abounded in Permian times, around 280 million years ago. There is only one living species of ginkgo. Its "primitive" fan-shaped leaves, whose veins branch in a series of Y-shapes, are identical to fossil leaves from Triassic rocks known to be 200 million years old. Ginkgos have been cultivated for centuries in China and Japan for their seeds, which are edible.

Monkey-puzzle, or *Araucaria,* trees are also living fossils. Fossil wood of similar structure has been found in Paleozoic rocks.

The first polluters

The oldest living fossils of all are found in Shark Bay, Australia. Strange rocky mounds up to 5 feet high grow in the shallow water here, often uncovered at low tide. They are made by blue-green algae, whose matted filaments trap sediments and somehow make the water deposit lime. The mounds, called stromatolites, are built of layers of algae and rocky cement.

Similar structures were widespread all over the globe in Precambrian times. In fact, fossils of almost identical stromatolites have been found in rocks 3 billion years old. The ancient stromatolites changed the world by releasing oxygen (from photosynthesis, see page 52) into

A forest of *Araucaria*, or monkey-puzzle, trees. These ancient conifers first appeared in Triassic times. Today they are found in South America, Australia, and New Guinea, a distribution that shows they evolved on the ancient supercontinent of Gondwanaland. These early seed-bearing plants produced their seeds on the undersides of the woody scale leaves that made up the cone (inset).

▲▼ Living (above) and fossil (below) leaves of the ginkgo. The ginkgo has survived almost unchanged for some 280 million years.

the atmosphere. This must have posed "pollution" problems for many existing living creatures, which had evolved to function in the absence of oxygen. However, new forms of life eventually evolved that used the oxygen to fuel a new and more efficient way of life, stimulating a great burst of evolution.

Most of the stromatolites died out about 80 million years ago. They may have been devastated by ice ages or other climatic changes, or perhaps they were eaten by early multicellular animals. Today, stromatolites are found in very few places. Shark Bay is a very special place. It is very hot, and there is very little rainfall and very little water movement. High rates of evaporation have left the water extremely salty – too salty for animals like snails and other shallow-water predators to survive. There must have been similar predator-free habitats somewhere in the world to allow stromatolites to survive for several billion years (see also page 52).

Last of the ammonites

Off the coast of the island of Vanuatu, in the Pacific Ocean, on dimly moonlit nights you may be lucky enough to see pale coiled shells bobbing in the water about 3 feet below the surface. From beneath the shells, large eyes peer out into the dark water; eyes that have seen strange creatures –ichthyosaurs, plesiosaurs, and armored fish – come and go, but have outlived them all. Normally deep-water animals, for some reason nautiluses rise to the surface here at certain times to hunt for lobsters and other shellfish, which they seize with their octopus-like tentacles. You could be fooled into thinking you were looking at a prehistoric sea of 200 million years ago.

The nautiluses are not ammonites, but they are close relatives that first appeared in the fossil record in the Ordovician period. Over 3,000 fossil species have been found, but today there are only six living species. Somehow they managed to survive the catastrophe that wiped out their relatives

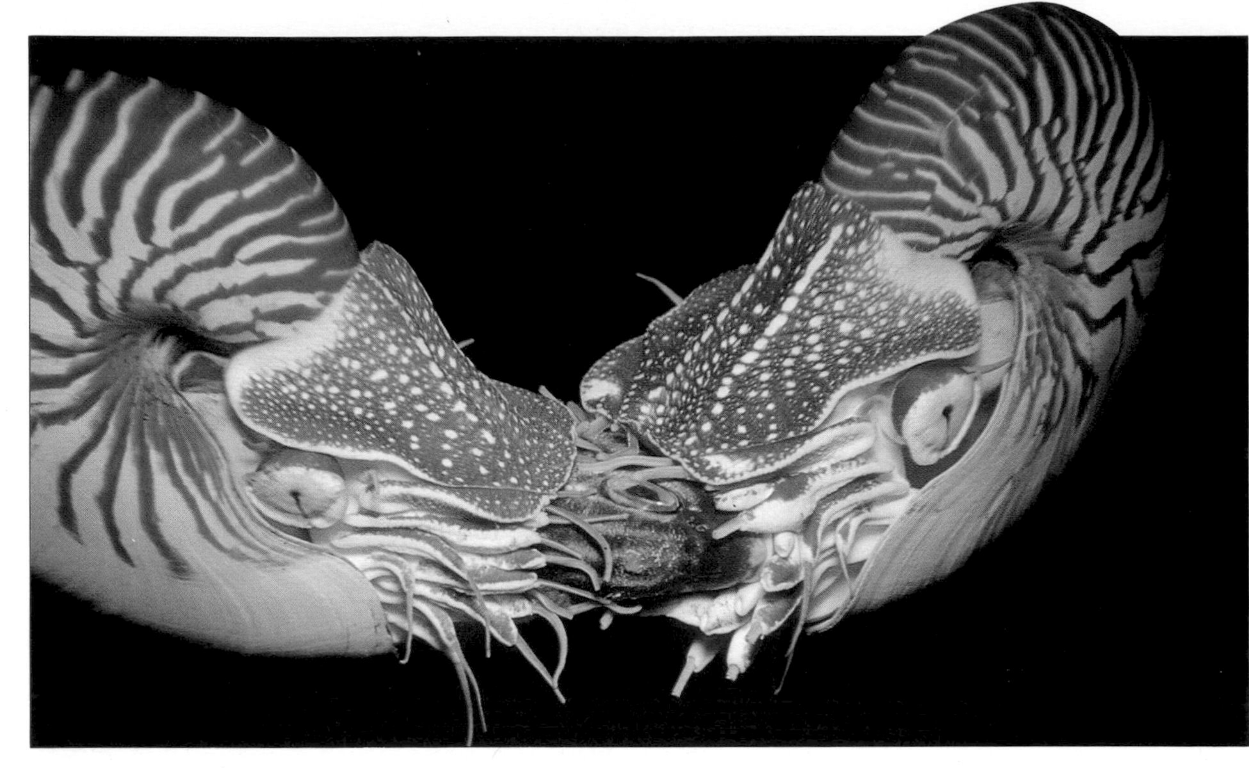

◀ Male and female nautiluses share a meal. Nautiluses are marine hunters related to squids and octopuses. Their shells are divided into chambers, some of which are filled with gas to help the animal float. The amount of gas can be adjusted if the nautilus wants to rise or sink in the water. In Ordovician times the ocean teemed with nautiloids, but after that they declined in numbers, and most became extinct.

the ammonites, the dinosaurs, and many other animals at the end of the Cretaceous period. Perhaps it was because they lived in deep water that they survived – the effects of the disaster may not have penetrated so deep into the oceans.

More deep-ocean survivors

A number of living fossils remained undiscovered for years, hidden in the depths of the oceans. They include some small mollusks that are very similar to some of the world's earliest mollusks – the five living species of monoplacophorans, or *Neopilina*. Until 1952 people thought the group became extinct some 400 million years ago. Then living *Neopilina* were found 12,300 feet down in a deep-sea trench in the Pacific Ocean. Since then, four other species have been found, all in deep-sea trenches. At first sight, monoplacophorans (the group to which *Neopolina* belong) resemble limpets; but inside the shell, organs like gills, nerve cords, excretory structures, and sex organs are arranged in pairs, rather like those of annelid worms and arthropods. This suggests that annelids, mollusks, and arthropods may have shared a common ancestor.

From the Pacific Ocean has come yet another possible living fossil, a descendant of a group that arose in the Cambrian period over 500 million years ago and was thought to have died out about 345 million years ago. This is a little animal called *Cephalodiscus*, known to scientists for many years. A new species was discovered in 1992, which looks remarkably like a graptolite, a small animal that establishes itself in its own "cup" linked by living connections to other cups. The individual *Cephalodiscus* animals hide in the cups by day and climb up spines on the cups at night to feed. Similar spines can be seen in many fossil graptolites.

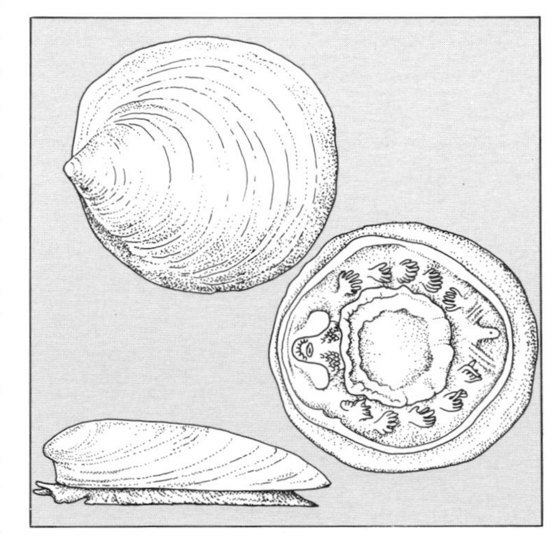

▲ In the deepest parts of the Pacific Ocean, several miles down, live tiny *Neopilina* mollusks that have probably remained unchanged for millions of years. Although they look like small limpets, their internal organs are paired and arranged rather like those of annelid worms and arthropods: They probably resemble the common ancestors of the worms, arthropods, and mollusks.

Monarch of an ancient kingdom

The horseshoe crab (king crab), which invades the beaches of North America in large numbers to spawn, is another survivor from ancient times. But this living fossil can hardly go unnoticed. Even a quick glance will reveal a creature that looks like a giant trilobite. Its young look even more like tiny trilobites.

The horseshoe crab is not a trilobite, but it belongs to an equally ancient group of animals, the arthropods, which first appeared about 550 million years ago. Horseshoe crabs have remained almost unchanged for at least 300 million years. They are remarkably adaptable and resilient. They can feed on almost any type of prey, and eat carrion (dead animals) as well.

Horseshoe crabs can cope with great swings in water temperature and saltiness, and with pollution. They can even survive for several days out of water. This means they can live in habitats like the shallow offshore waters and the seashore, where harsh conditions limit the number of predators. In any case, their hard armor-plated heads and bodies present a tough prospect for potential enemies.

Clues to the past

Horseshoe crabs live mainly on the seabed, but they can swim by paddling, lying on their backs. By watching how horseshoe crabs move through sand and mud, scientists have learned the meaning of many tracks and traces in ancient sedimentary rocks. The crabs plow their way through the upper sediments, using their tail spine to lever themselves down, and one pair of walking legs as shovels. Their other walking legs have jawlike pincers with which to seize prey and crack open shells.

Beware of dragons!

Imagine a small dragon, about 25 inches long, with a fearsome array of jagged horny plates along its back. The date is 140 million years ago, and the dinosaurs are still around. By modern reptile standards it is primitive. Its heart is very primitive, and the bones of its skull are arranged like those of crocodiles and dinosaurs, but not modern lizards. However, like modern lizards, it has a third "eye," a light-sensitive structure under the skin on the top of its head.

Now step forward into the present day, to the cliff tops of some islands off the New Zealand coast. An identical dragon is basking in the sun at the entrance to a seabird's burrow, which it has taken over. This is the tuatara, the only survivor of a group of reptiles that arose over 200 million years ago. It is nocturnal (active at night), and its eyes have a bright reflective layer to help it see in dim light. It feeds on insects, worms, slugs, and snails, the chicks and eggs of the seabirds among which it lives, and even on smaller tuataras.

Why has it survived? New Zealand became isolated from the rest of the world about 80 million years ago, before many modern-day predators had invaded the islands. Although it lays no more than 15 eggs at a time, and these take some 15 months to incubate, the tuatara probably lives for over 120 years, so it has plenty of opportunities to reproduce.

How important are living fossils?

Living fossils offer fascinating glimpses of a long vanished world. They can also sometimes give valuable information about ancient groups of animals. Some, like *Neopilina* and the tuatara, may perhaps provide clues to the links between groups of animals – the mollusks and the annelid worms, the lizards and the dinosaurs. Others, like the coelacanth, are more of a puzzle. Are they missing links or evolutionary blind alleys?

▼ Safe in the isolation of New Zealand, the tuatara is a survivor from the age of dinosaurs. It is the sole living member of an ancient order of reptiles with certain primitive skeletal features, a primitive heart, no eardrum, and no true teeth – just serrations of the jawbone.

◀ Horseshoe crabs spawning on a beach in New Jersey. More closely related to spiders and scorpions than to crabs, these creatures have survived virtually unchanged for over 300 million years: This could almost be a picture of a beach in Carboniferous times.

The Geological Record

Rocks and the fossils they contain tell the story of Earth, its changing climate and wandering continents, the evolution of its atmosphere and its mountain ranges, its seas and lakes, and, above all, its life.

The record of life preserved in fossils – the fossil record – shows us the changing complexity and diversity of animal and plant life. Sometimes it is possible to trace the gradual changes in the bodies of particular groups of animals, to see how new groups of animals have emerged. This process of change is called evolution.

When we talk about evolution, we are usually talking about plant and animal species. But what is a species? The simplest explanation is that members of the same species will breed together and produce offspring, and these offspring in turn can also breed together successfully. Members of different species cannot produce fertile offspring, and usually they will not even try to mate with each other. Perhaps they have different courtship displays, so they do not send each other the right signals. Or they may breed in different seasons. Often they simply live in different places (perhaps cut off by continental drift or by new mountain ranges), so they never meet.

Species that come and go

There are about 1.5 million known species today, and probably over 30 million still to be discovered (a large proportion of which will be insects). Some of these species have existed for only a few hundred years, while others have been around for hundreds of millions of years. How long a species lasts depends on how fast that group of animals is evolving, and whether it is competing with other species for valuable food or space. Mussels and clams, for example, are evolving relatively slowly. On the average a mussel species survives for 50 million years. A mammal species, on the other hand, usually lasts only 5 million years.

What's in a name?

Prehistoric monsters such as the dinosaurs seem even more exotic because of their strange names. The name *dinosaur* itself means "terrible lizard." *Tyrannosaurus rex* means "king of the tyrant reptiles," an apt name for such a fearsome monster, one of the greatest hunters ever to stalk the earth. The flying reptile *Pterodactylus* had wings (*ptero*) supported by the bones of its fingers (*dactylus*), while the ichthyosaurs were fishlike (*icthyo*) lizards (*saurus*). The primitive seed plant *Glossopteris* is a fern (*pteris*) with tongue-shaped (*glosso*) leaves.

Sometimes species are named after the place where their fossils were first discovered. *Mesosaurus*, for instance, was first found near the river Meuse. Or they may be named in honor of the person who discovered them. *Yaleosaurus* was named after Yale University, the university of its discoverer.

These names are Latin names or names with Latin endings. Each different kind of creature – each species – has two special Latin names. The first name – *Tyrannosaurus*, for example – is the genus to which it belongs; *rex* is the species name. Similar species are grouped together in the same genus. Animals or plants of different species that belong to the same genus are more alike than animals or plants that belong to different genera. Your pet dog is called *Canis vulgaris* – "dog common" – while the gray wolf, which is also a kind of dog, is called *Canis lupus*. In Latin, the adjective comes after the noun, so instead of saying "common dog," you say "dog common."

Scientists use Latin names as a kind of international language. The word for *cat* or *dog* is different in every language, but if everyone uses the Latin name, we know they are speaking about the same creature. We also know whether they are talking about a domestic cat or a wildcat, a bobcat or a lynx. The system of giving two names to every species is called the binomial system (Latin *bi* = two, *nom* = name); it was first proposed by the Swedish biologist Linnaeus (Carl von Linné) in 1735.

Close, closer, closest

There are groups bigger than genera (plural of genus). Many different genera make up a family. Dogs belong to the family Canidae, the dog family. This family and many others belong to the order Carnivora (the carnivorous or flesh-eating mammals). This includes such animals as bears, dogs, cats, and weasels. These belong to the class Mammalia (the mammals) in the phylum Chordata (animals with a nerve cord running along their backs and a segmented supporting rod – in vertebrates, the backbone – at some stage in their life cycles); they include the fish, amphibians, reptiles, birds, and mammals. These are included in the kingdom Animalia (the animals). Grouping animals like this makes it easy to see which are closely related. The further down the chart the animals come, the more features they have in common.

A living calendar

The fossil record shows many examples of new species appearing and old ones disappearing. Sometimes particular groups of animals change very rapidly. Perhaps they are moving into new, unoccupied habitats, or maybe their climate or surroundings are changing so they have to adapt or die. Fossils of animal groups that change quickly over time can be used to date rocks. They are called index fossils. Rocks with the same index fossils are likely to be of the same age.

Trilobites are important index fossils for sequencing Cambrian rocks; in Ordovician and Silurian times the brachiopods and graptolites changed rapidly. For many Devonian to Cretaceous

THE CLASSIFICATION OF *TYRANNOSAURUS REX*
Species – *Tyrannosaurus rex*
Genus – *Tyrannosaurus*
Albertosaurus 30 feet
Alioramus 20 feet
Daspletosaurus 30 feet
Tarbosaurus 40 feet
Tyrannosaurus 44 feet
Family – Tyrannosauridae
Suborder – Therapoda (flesh eaters walking on two legs)
Order – Saurischia (lizard-hipped dinosaurs)
Class – Reptilia (reptiles)
Phylum – Chordata (animals with notochord and dorsal nerve chord)
Kingdom – Animalia (animals)

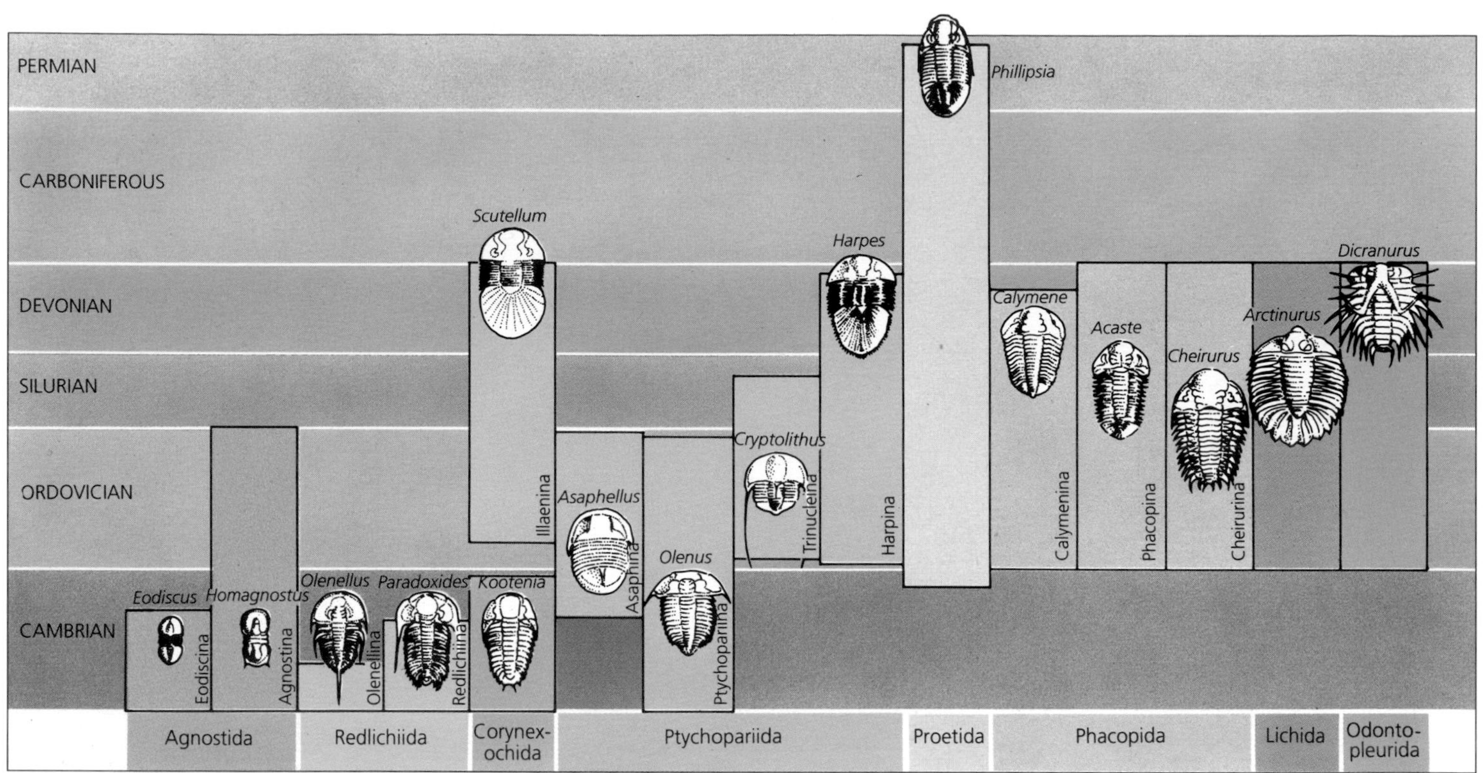

▲ Trilobites are useful index fossils. Certain species and genera occurred for only a short period of time, so their presence indicates the age of rocks. This chart shows the most important groups of trilobites. Individual species had shorter life spans.

rocks ammonites are used; and for the Jurassic and Cretaceous periods the microscopic shells of single-celled foraminifers are extremely important index fossils.

Gaps in the record

All too often there are gaps in the record. Perhaps no sediments were deposited in the places where the missing animals lived. Or maybe these sediments have long since been altered beyond recognition by the great forces of mountain building or been destroyed by erosion.

One of the biggest gaps is between the traces and occasional imprints of soft-bodied animals in Precambrian rocks, and the abundant trilobites and other hard-shelled animals in early Cambrian rocks, a few million years later. Until they had evolved calcium carbonate (lime) skeletons, they were not readily fossilized. Another mystery is the origin of the fish – the first group of vertebrates to evolve.

Missing links

There are probably several different reasons for the missing links. One problem is that when groups of animals are evolving, many of the species that arise are not very successful and do not survive for long, so there is very little chance of any of them becoming fossils. Some scientists think that from time to time there are rapid bursts of evolution in particular animal groups. These produce a succession of short-lived species, very few of which are around long enough to have a chance of being fossilized.

The geological time scale

The history of the earth has been divided into geological periods – the geological time scale. The major periods generally correspond to marked changes in the earth's climate or fossil record. Their boundaries can be recognized from their fossils, and also from the type of rocks being laid down. The Triassic period, for example, saw a change to hot, dry conditions, with lots of red oxidized sandstones, while the Cretaceous period produced huge thicknesses of chalk. The Cenozoic periods are based on the number of mollusk fossils in the rocks.

In turn, these periods have been grouped into eras or aeons. There are five eras. The oldest and longest is the Archaean, the period before life appeared. The Proterozoic ("earlier life") is marked by an absence of hard-bodied fossils. Then comes the Paleozoic ("ancient life"), which starts with the appearance of shelled animals in the fossil record and ends with the great extinction of the late Permian period. The Mesozoic ("middle life") ends with the disappearance of the dinosaurs in another mass extinction at the end of the Cretaceous period, and the Cenozoic ("new life") continues to this day. (The geolocial time scale is illustrated on pages 152-153.)

THE MEANINGS OF SOME DINOSAUR NAMES

Iguanodon – iguana tooth (the first part to be found was a tooth, which was very like an iguana tooth)

Deinodon horridus – horrible terror tooth

Deinocheirus – terrible hand

Deinonychus – terrible claw

Oviraptor – egg stealer

Velociraptor – swift robber

Triceratops – three-horned face

Ornithomimus – bird mimic

Psittacosaurus – parrot lizard

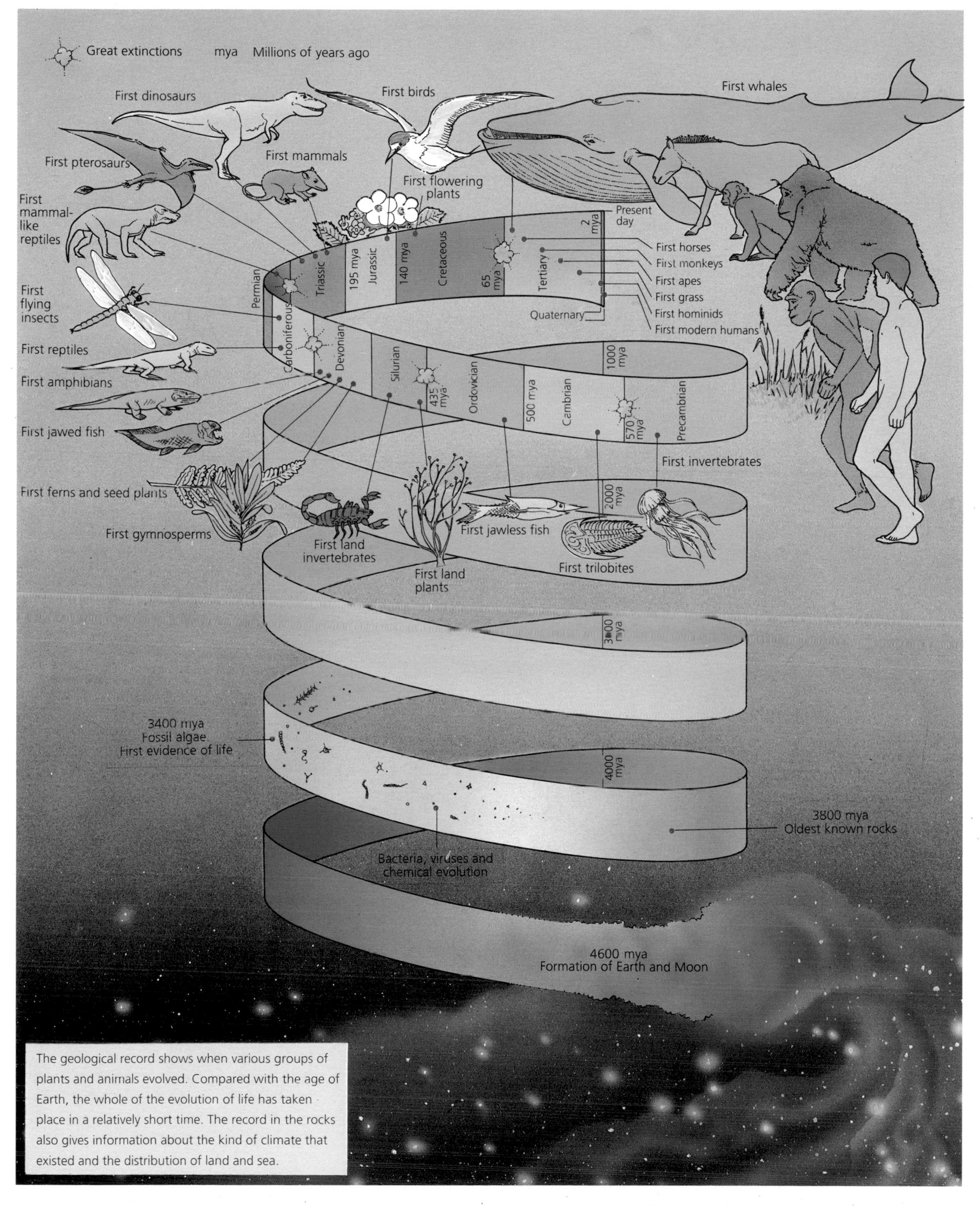

The geological record shows when various groups of plants and animals evolved. Compared with the age of Earth, the whole of the evolution of life has taken place in a relatively short time. The record in the rocks also gives information about the kind of climate that existed and the distribution of land and sea.

THE EVOLUTION OF LIFE

Different animals in the fossil record appeared (and often disappeared) at different times. The very tiny changes that occurred in animals gradually produced, after millions of years, quite different animals. These gradual changes in living things as time goes on are called evolution. If you assume that similar animals had a common ancestor, you can plot a kind of family tree of the animal kingdom.

In the middle of the 19th century two men, Charles Darwin and Alfred Russel Wallace, proposed the theory of evolution to explain how new species arise from species that already exist. They said that when animals or plants are under pressure, perhaps from food shortage, overcrowding, or predators, only certain individuals will manage to survive long enough to breed. These are the "fittest" individuals – the ones best adapted to the local conditions. They have certain inherited features that make them more successful than their neighbors. The offspring of the "fittest" individuals may also have these successful features, so there will be a greater proportion of these features in the next generation than before. So the most successful features will gradually spread through a population.

The natural pressures to which living things must adapt are the driving force that produces evolution. Darwin called this process natural selection.

Variation – the key to evolution

Evolution can work only if there is plenty of variation in the population. So long as there are individuals with a wide range of features, some will be better adapted to local conditions than others. There are two kinds of variation – inherited and non-inherited variation. Non-inherited variation is due to a creature's surroundings – whether an animal has enough food, or a plant has enough light as it is growing up, for example. Inherited variation depends on characteristics passed down from generation to generation. You can see this in human populations. Someone whose parents are both very short is unlikely to grow very tall, but exactly how tall he or she grows will also depend on the quality of the food eaten as a child.

Changing the program

Everyone knows families in which brothers and sisters look very different from each other, even though they have the same parents. How does this happen? Inherited characteristics are produced by a special code called the genetic code, contained in a chemical (called DNA) that is present in every cell. They are passed on when the DNA of a sperm joins with the DNA of an egg during reproduction. This mixing up of the DNA from two different parents leads to new combinations of characteristics in the offspring.

THE RISE OF THE PTEROSAURS – AN EXAMPLE OF EVOLUTION

During the Mesozoic era, 248 to 65 million years ago, a group of bird-like reptiles evolved the ability to fly. They were the pterosaurs. Scientists have tried to piece together the fossil clues and determine how this transformation took place. They can only guess at the missing evidence, but they have suggested a sequence of changes, each of which might have given the animal some new advantage. The diagrams show how many small changes in the pterosaur's body eventually produced an entirely new sort of animal.

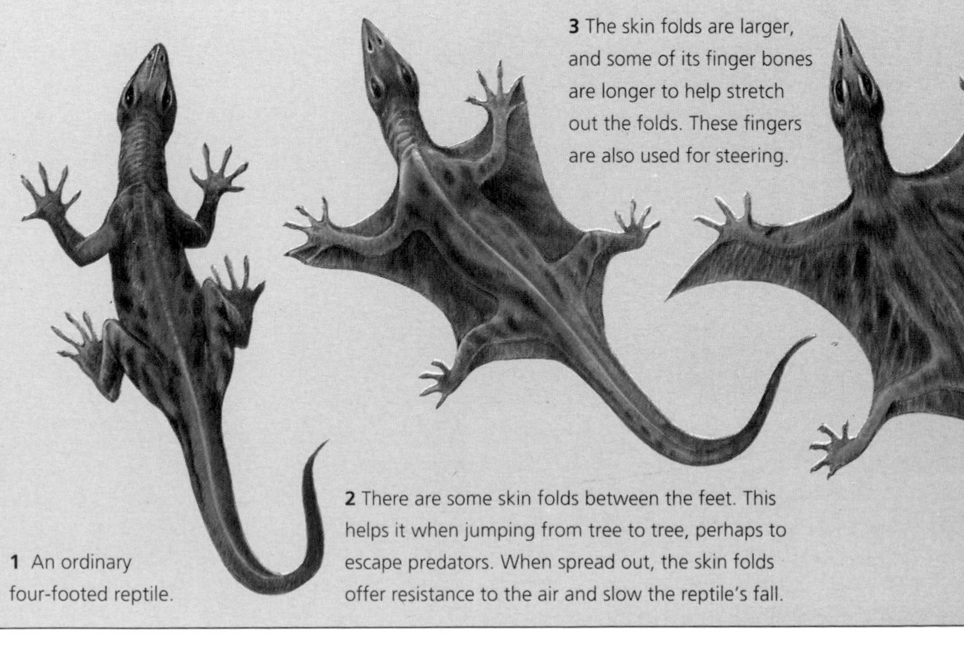

1 An ordinary four-footed reptile.

2 There are some skin folds between the feet. This helps it when jumping from tree to tree, perhaps to escape predators. When spread out, the skin folds offer resistance to the air and slow the reptile's fall.

3 The skin folds are larger, and some of its finger bones are longer to help stretch out the folds. These fingers are also used for steering.

On rare occasions, variation is due to changes in the DNA itself. Harmful radiation or chemicals taken into the body can change the DNA. Or cells may make mistakes when copying their DNA as they divide. In these ways completely new characteristics may arise.

Growing apart

New species often arise in response to changes in the earth's surface and climate. When mountain ranges rise up or rivers change course, different populations of the same species become cut off from each other. Over many generations, these populations become more and more specialized – adapted to their own special local conditions. These conditions may become very different when continents drift apart and change latitude, so that the climate changes drastically. Eventually the two populations become so different that they can no longer breed together: They are now separate species.

Living history books

The history of the continents can be deduced from the geographical distribution of certain groups of related animals that are found in different parts of the world today. Among the strangest are the large flightless birds: the African ostriches, South American rheas, Australian emus, and giant moas of New Zealand (which are now extinct). Such birds are found nowhere else in the world. This is because they first evolved when Africa, Australia, and South America were joined together in the great southern supercontinent of Gondwanaland. They evolved after the northern continents had separated from Gondwanaland, so they are not found in Europe and North America.

Copycats?

Over millions of years plants and animals on islands, and on continents, evolve in their own special way. Australia and New Guinea, for example, have a unique collection of animals. These are a kind of mammal called marsupials, which give birth to very undeveloped young and carry them in pouches on their bellies. Australia and New Guinea have been separated from the other continents for a very long time, and evolution has taken a different course here. In the rest of the world, most mammals are placental mammals, which carry their young inside them, attached by a placenta, until they are very well developed.

There are many different kinds of marsupials: slow-moving koalas that feed on tough leaves just like sloths, burrowing bandicoots that behave like rabbits, and fierce marsupial cats that hunt marsupial mice. There were even, until fairly recently, a marsupial wolf and a marsupial lion. The marsupials have evolved species that are similar to placental mammal species that occupy similar habitats. The way unrelated plants and animals from different parts of the world tend to evolve similar features when adapting to similar environments is called convergent evolution.

Even among the placental mammals, you can see convergent evolution at work. There are anteaters in South and Central America with huge curving claws on their front paws for tearing at ant and termite nests, long snouts, and long sticky tongues for licking up their prey. In Africa there are no anteaters but there are ant-eating aardvarks and pangolins with similar adaptations. Plants can be copycats, too. The American deserts are full of fat, succulent cacti, armed with spines, while the African deserts have succulent, spiny euphorbias.

A procession of species

Animals and plants may become isolated on islands when the sea level rises or the continents move apart. Islands are often invaded by accident: Birds get blown off course during migration; reptiles and insects accidentally get carried across the oceans on rafts of floating vegetation; seeds get blown by the wind or carried on the feet of birds. Even new volcanic islands are soon covered in green vegetation that supports a large population of birds, insects, and other animals.

4 The wing fingers are much longer and thicker, and the wing membrane is now so large that it can be flapped. The breastbones are larger and stronger to support the muscles needed for flying. The back legs have also changed position, to make the body more streamlined when flying.

5 A fully evolved pterosaur The wing fingers are even larger, and the tail has a flat membrane at the tip for steering and balancing.

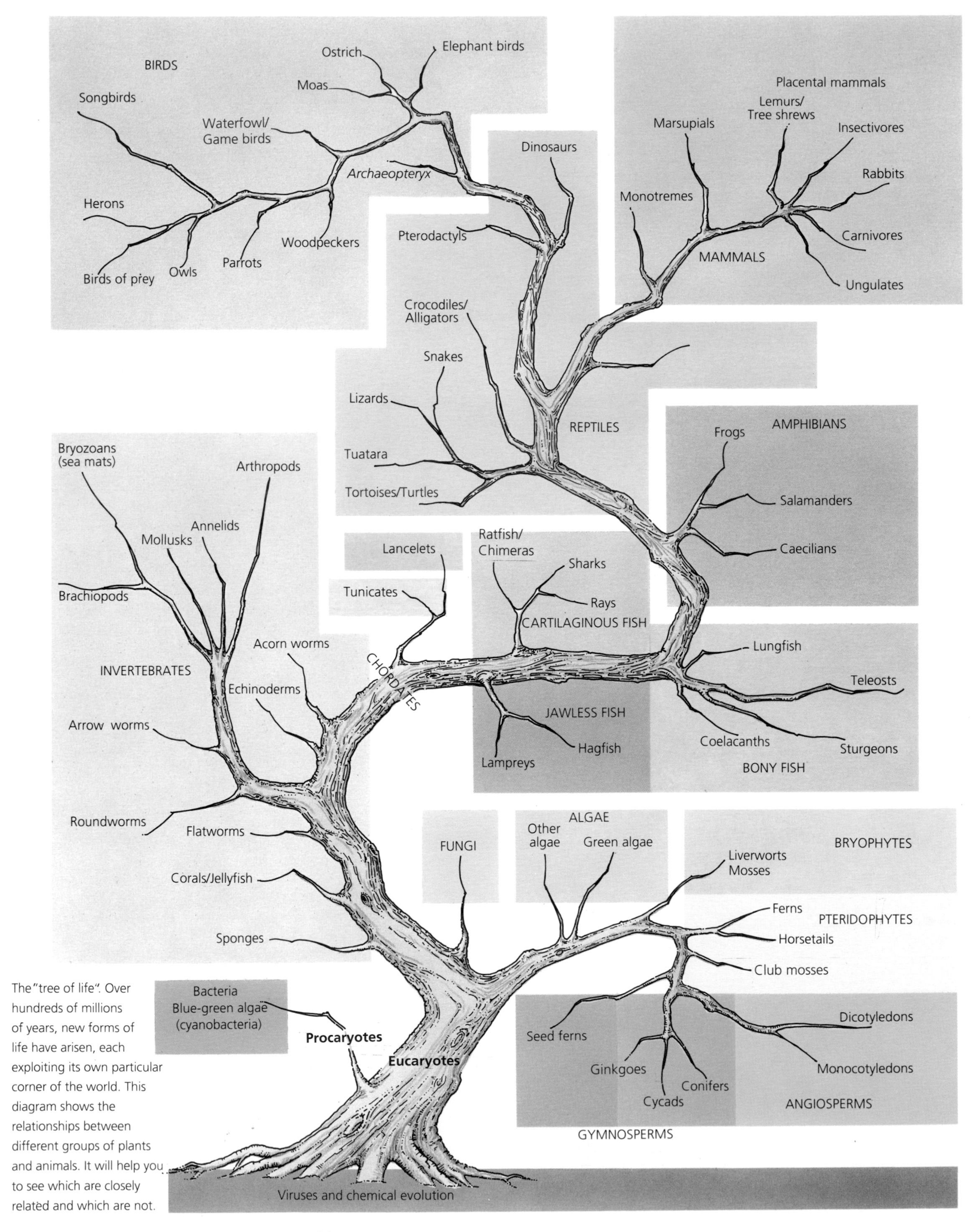

The "tree of life". Over hundreds of millions of years, new forms of life have arisen, each exploiting its own particular corner of the world. This diagram shows the relationships between different groups of plants and animals. It will help you to see which are closely related and which are not.

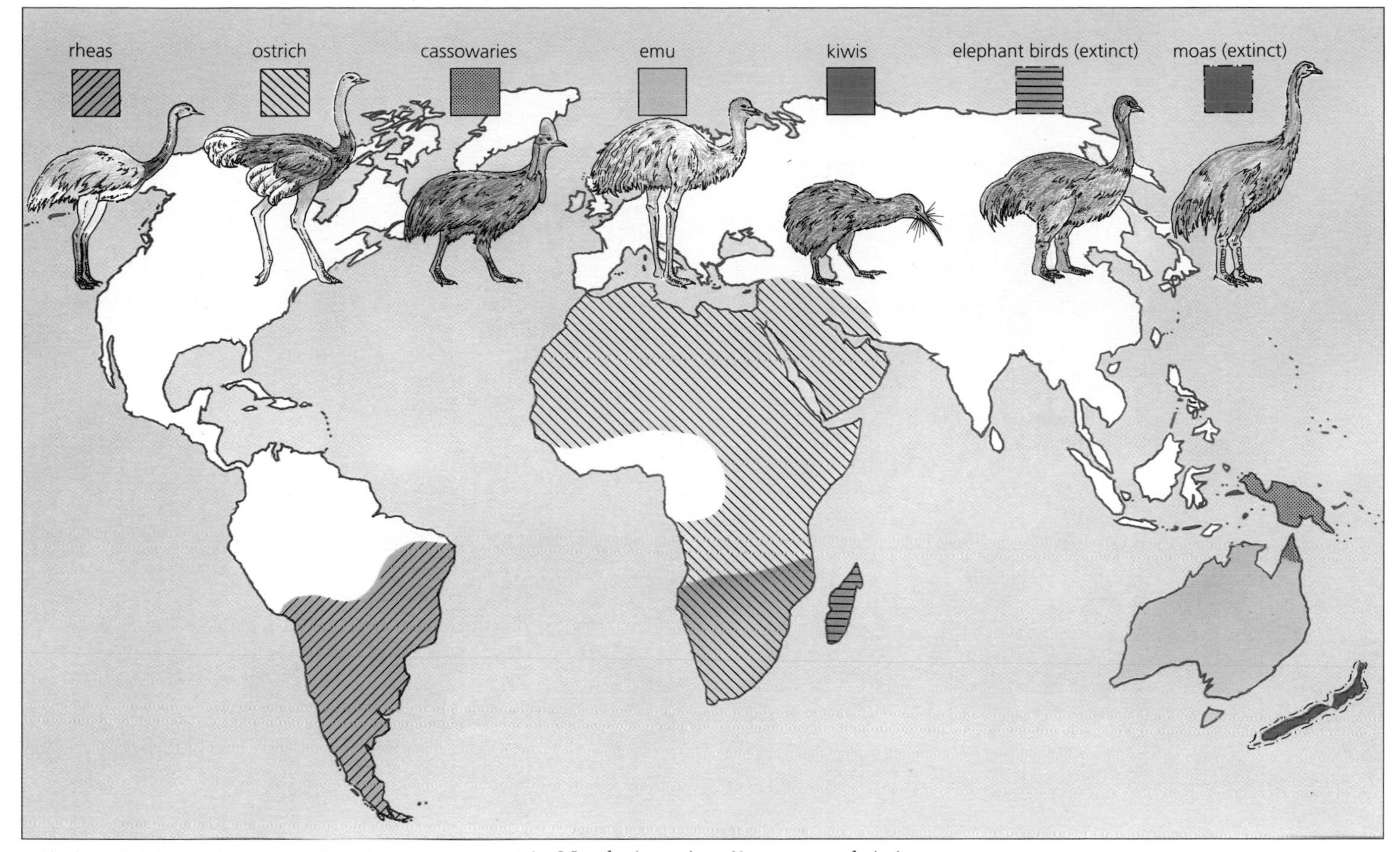

▲ The large flightless birds known as the ratites are not found in Europe or North America because their ancestors evolved on the supercontinent of Gondwanaland after it had separated from the northern continents.

The animals that arrive on islands may find very little competition, and many empty habitats to invade. Species can often thrive on islands, whereas they would be held back by competition from other species on the mainland. Once populations build up, competition increases. One solution is for an animal to become more specialized – to eat a food no other animal wants, for instance. Many new species evolve from the few that first arrived on the island. This process is called adaptive radiation. A similar thing occurs when new habitats appear, perhaps as the sea retreats from the land or new mountain ranges rise up.

No teeth for change

At one time marsupial (pouched) mammals were common on all the continents, but as placental mammal carnivores evolved, most of the marsupials became extinct. By 25 million years ago, marsupials had died out in North America, Europe, and Asia. The reason for this is a mystery. There seems to be no obvious reason why placental predators should be more successful hunters than marsupial predators. But perhaps it was not a simple question of competition. When the mammals were evolving, the continents were drifting into new parts of the world and their climates were changing. It may be that the placental carnivores were more adaptable.

One suggestion is that the answer lies in the teeth. Meat eaters need sharp scissorlike teeth, called carnassials, to slice up their prey. The cheek teeth of marsupial carnivores were all sharp and scissorlike, but the placental carnivores also had several other types of teeth in their jaws, as humans do. If the climate changed or prey was scarce, the placental mammals could

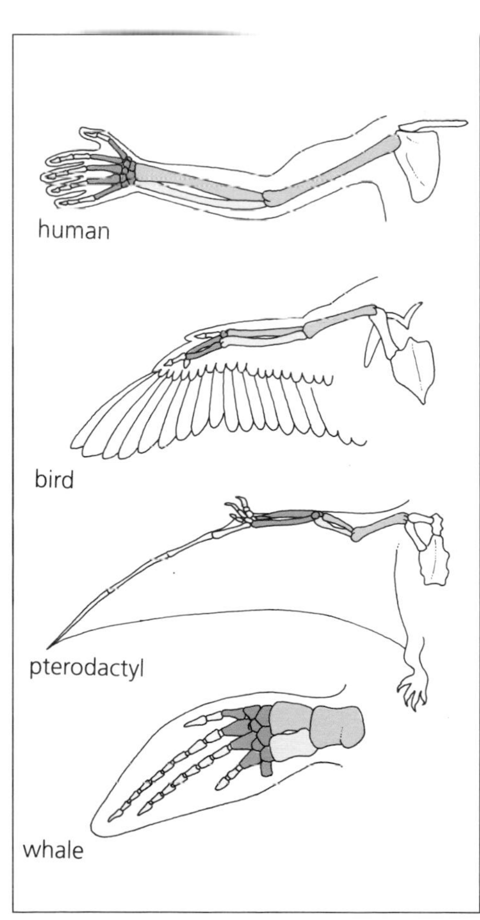

► Comparing the bodily structures of animals can provide clues to their ancestry. For example, the forelimbs of vertebrates are all made up of the same bones in the same basic arrangement, which suggests that they share a common ancestor. Structures that appear dissimilar but have similar anatomy are called homologous structures.

◄ ► Animals that live in different geographical regions and are unrelated often evolve similar adaptations if they live in similar habitats or have similar life-styles. This is called convergent evolution. The koala (right), an Australian marsupial, and the sloth (left), a placental mammal from South and Central America, both have a very slow, lazy life-style. They move slowly, and they feed exclusively on a limited variety of leaves that are not particularly nutritious. Their slow life-style means that they do not use much energy, so they can manage with less food than other mammals. By feeding on leaves that other animals do not want, they avoid competition for food.

probably adapt their diets and survive. Some modern carnivores, such as brown bears, often kill their own prey, but they also feed on fish or berries at certain seasons. Maybe marsupial carnivores had teeth that were only good for meat eating – you cannot grind very well with sharp pointed teeth.

Protected from progress

Long after the marsupials were wiped out from the northern continents, they still thrived in South America, Antarctica, Australia, and New Guinea. These continents were all joined together in a vast southern continent, Gondwanaland, which later broke up. South America drifted toward North America, while Antarctica moved over the South Pole and became very cold, and Australia and New Guinea drifted toward the equator.

About 2 million years ago, a chain of volcanoes arose between North and South America, forming first stepping-stone islands, then a real land bridge between the two continents. This led to a mixing of the animals of North and South America. The result was a decline in the marsupials of South America, and most became extinct. A few survive to this day, and some have invaded North America. The omnivorous Virginia opossum is still advancing northward through North America, and seems to be very successful.

In Australia and New Guinea, already cut off from the other continents, the marsupials survived. But today Australia's marsupials are in great danger. Instead of a land bridge there is a human bridge: Humans have brought in placental mammals such as goats, rabbits, dogs, and cats, and many marsupial species are in danger of being wiped out.

The cost of progress

As new, more efficient kinds of animals evolve, some older species cannot compete, and go into a decline. In the oceans, the earliest predators were the trilobites. Then the nautiloids evolved. They were faster swimmers, and their powerful jaws could crush trilobite shells. Then for a time the ammonoids became more successful than the nautiloids. They were badly affected by several mass extinctions, after which they evolved rapidly again, until finally they became extinct at the end of the Cretaceous period. The nautiloids survived and began to increase again after the decline of the ammonoids, but by then they were severely outnumbered by the bony fish. Today the fish are the major predators. They have evolved into thousands of different species that between them exploit almost every corner of the oceans.

Evolution and extinction

There are many reasons why species become extinct. They may suffer from competition or predation from newly evolved species (including humans) that are more advanced and more efficient. Continents may drift together, bringing different groups of animals into competition. Or a species's environment may change faster than it can adapt to it. If a species has become very specialized, the total variation in all the members of the species may not be enough to allow it to adapt to a change in its environment. Compare the success of the red fox with the decline of the giant panda. The fox eats anything from earthworms to rabbits, but the panda eats almost nothing but bamboo.

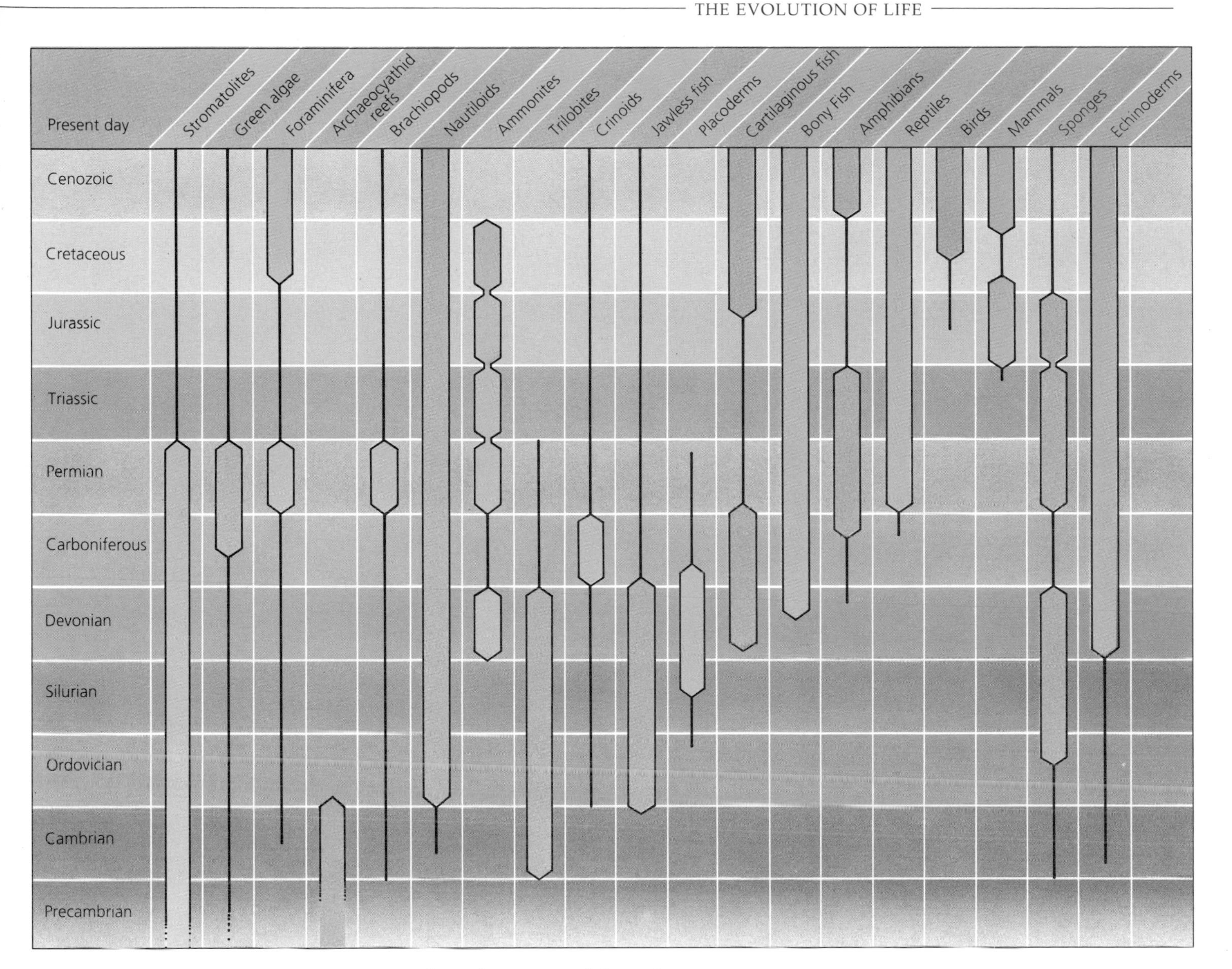

▲ Throughout the history of life, species have come and gone, and new species have arisen in their place. This chart shows the history of some of the major groups of animals. Relatively rapid changes in the distribution of land and sea or in the climate, such as the arrival or end of an ice age with its accompanying changes in sea level, have at times driven many species to extinction in a very short period of geological time. These mass extinctions are usually followed by a burst of new evolution, as existing species adapt to the new conditions.

The great extinctions

During the earth's history there have been several periods when huge numbers of species have become extinct. Probably the greatest mass extinction occurred about 250 million years ago, at the end of the Permian period, when 76 to 96 percent of all the species on earth disappeared – about 200 out of the 400 known families. Another mass extinction occurred at the end of the Cretaceous period, 65 million years ago, when the dinosaurs and ammonites vanished.

The causes of mass extinctions remain a big mystery, and many explanations have been put forward. Global climate change is the most likely reason, perhaps related to movements of the continents or to changes in the tilt of the earth's axis. It is possible that the great Cretaceous extinction may have been caused by a huge meteorite hitting the earth. This would have thrown so much dust into the atmosphere that the climate would have cooled very suddenly. Nitrogen oxides formed in the high temperature of the explosion would have dissolved in rain to give very acid rainfall, damaging plant life.

Just because an animal has become extinct does not mean that it was less efficient or less "advanced" than modern animals. Even the best modern predators, such as lions or bears, are not so well equipped as the dinosaurs were for slashing open the tough hides of the reptiles and amphibians that roamed the land in Cretaceous times.

Each major extinction has been followed by a burst of evolution, as new creatures have evolved to fill the now empty habitats and take the place of the vanished species.

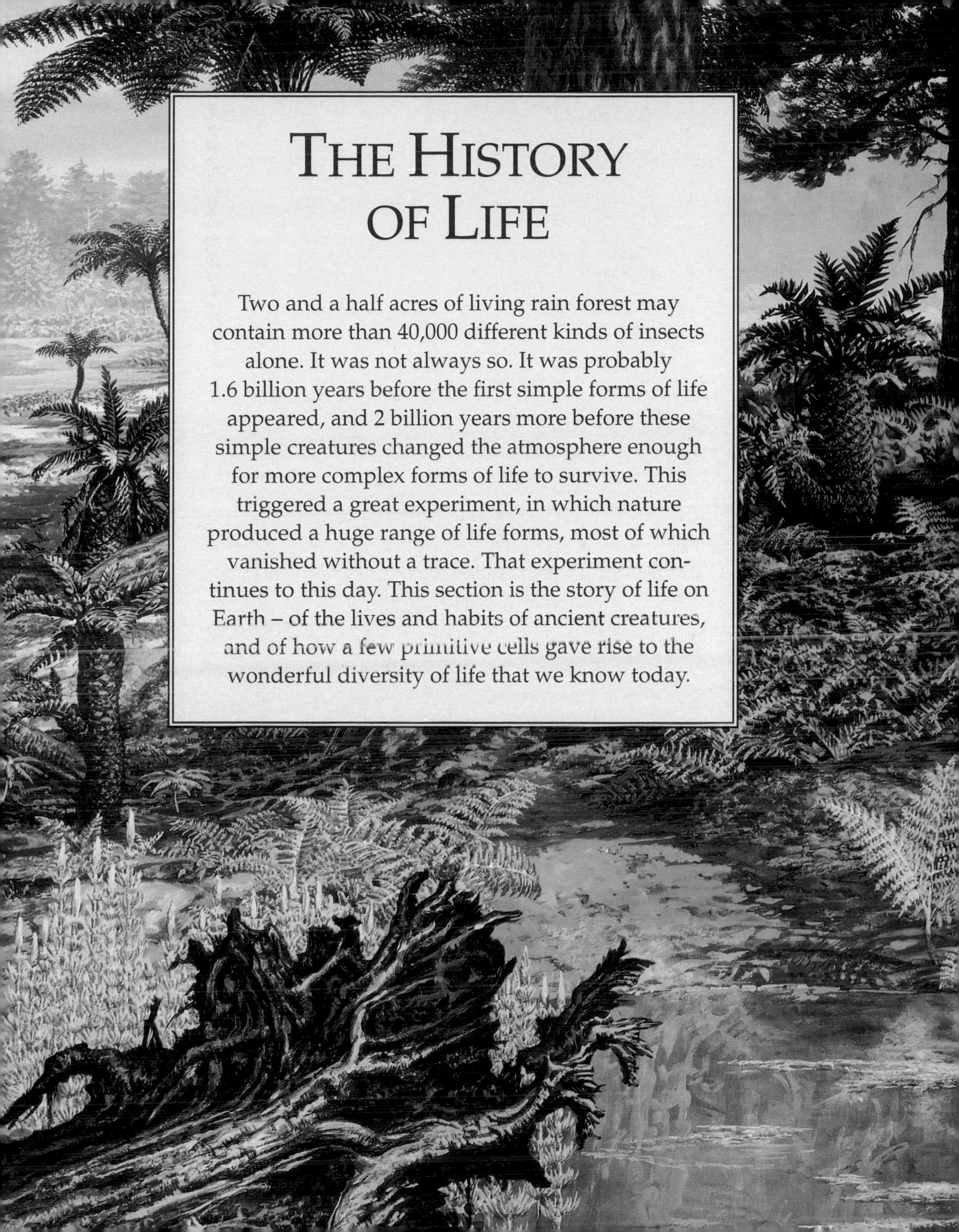

THE HISTORY OF LIFE

Two and a half acres of living rain forest may contain more than 40,000 different kinds of insects alone. It was not always so. It was probably 1.6 billion years before the first simple forms of life appeared, and 2 billion years more before these simple creatures changed the atmosphere enough for more complex forms of life to survive. This triggered a great experiment, in which nature produced a huge range of life forms, most of which vanished without a trace. That experiment continues to this day. This section is the story of life on Earth – of the lives and habits of ancient creatures, and of how a few primitive cells gave rise to the wonderful diversity of life that we know today.

The Precambrian Period

FORMATION OF EARTH TO 570 MILLION YEARS AGO

The Precambrian period stretches from the birth of Earth to the appearance of many-celled animals about 570 million years ago. The oldest-known rocks date back only 3.9 billion years, so we know very little about the early years of Earth. Even these rocks are so altered that they do not tell us very much.

Around 2.5 billion years ago, there was probably one large supercontinent, which later broke up into several smaller continents. By the end of the Precambrian period the continents had come together again to form a new supercontinent. Along with all these changes in land and sea came great changes in climate. There were at least three ice ages during the Precambrian, the oldest at about 2.3 billion years ago. The greatest ice age took place between 1 billion and 600 million years ago.

The early atmosphere did not contain oxygen gas. It was made up mainly of the gases methane and ammonia, with smaller amounts of hydrogen sulphide, water vapor, nitrogen, hydrogen, carbon monoxide, and carbon dioxide. But all this was to change when life finally appeared on Earth.

What is life?

Although today there are many different kinds of living creatures, they all have something in common. They are all made up of tiny baglike structures called cells. The bag is made of a very thin sheet called a membrane, which separates the chemicals inside the cell from its surroundings. The cells take in materials from the world around them and use them to grow. Sooner or later the living creature multiplies.

The stuff of life

Living creatures all contain certain special chemical compounds. Most of the cell's structures are either made of proteins or made by proteins. All the proteins found in living things are made of strings of chemicals called amino acids. All cells also contain another chemical, called ATP, that they use to store energy. The blueprint for making more cells – and even a new animal or plant – takes the form of a chemical code, in a long molecule called DNA. Each different kind of creature has its own special kind of DNA. Proteins, ATP, and DNA all contain carbon – they are all organic compounds. But how were all these chemicals produced in the first place?

Experimenting with life

The gases of the early atmosphere dissolved in the water of the oceans, resulting in a warm soup of chemicals. Without oxygen, there was no ozone layer (ozone is a form of oxygen) to protect the earth's surface from harmful, high-energy ultraviolet rays coming from the sun. In the 1920s, Russian scientist Alexandr Oparin and English scientist John Haldane suggested that over millions of years these rays, together with lightning discharges, could act on the chemical soup to form more complex chemicals, until eventually one was formed – DNA – that was able to make copies of itself.

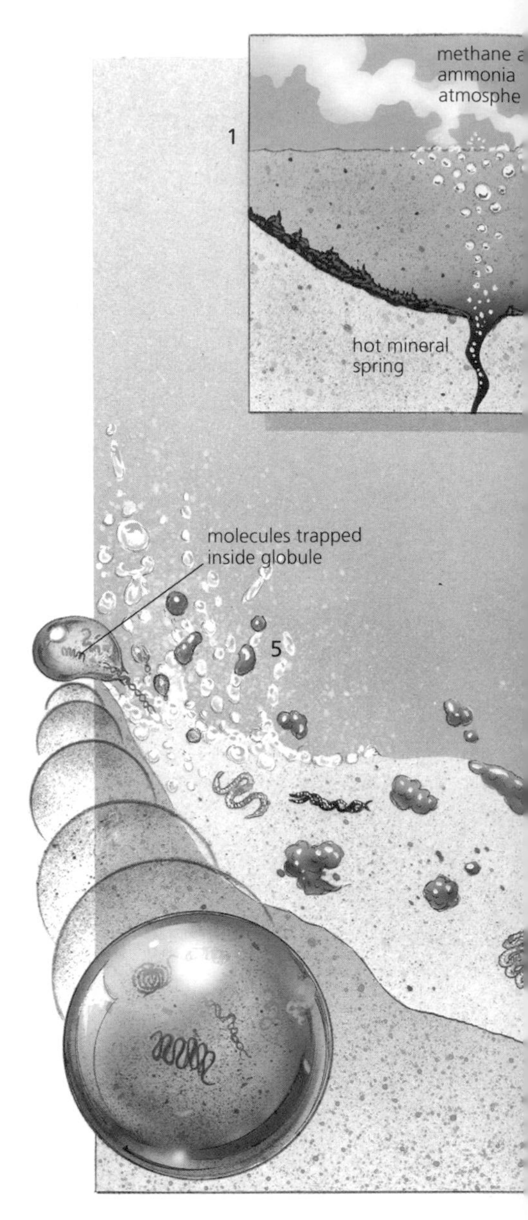

▲ The first cells. Methane and other gases in the primitive atmosphere of the earth dissolved in the water of seas, lakes, and pools to form a complex chemical "soup." Laboratory experiments have shown that lightning discharges acting on such a soup can cause the chemicals to react together and produce more complex chemicals very similar to those found in living cells. Eventually some of these chemicals developed the ability to reproduce themselves – to make copies of themselves. In the same soup there were also fat globules. If the soup was stirred violently by the wind, the complex chemicals may have become trapped inside fat globules. In time these hybrid structures evolved into living cells surrounded by fatty membranes.

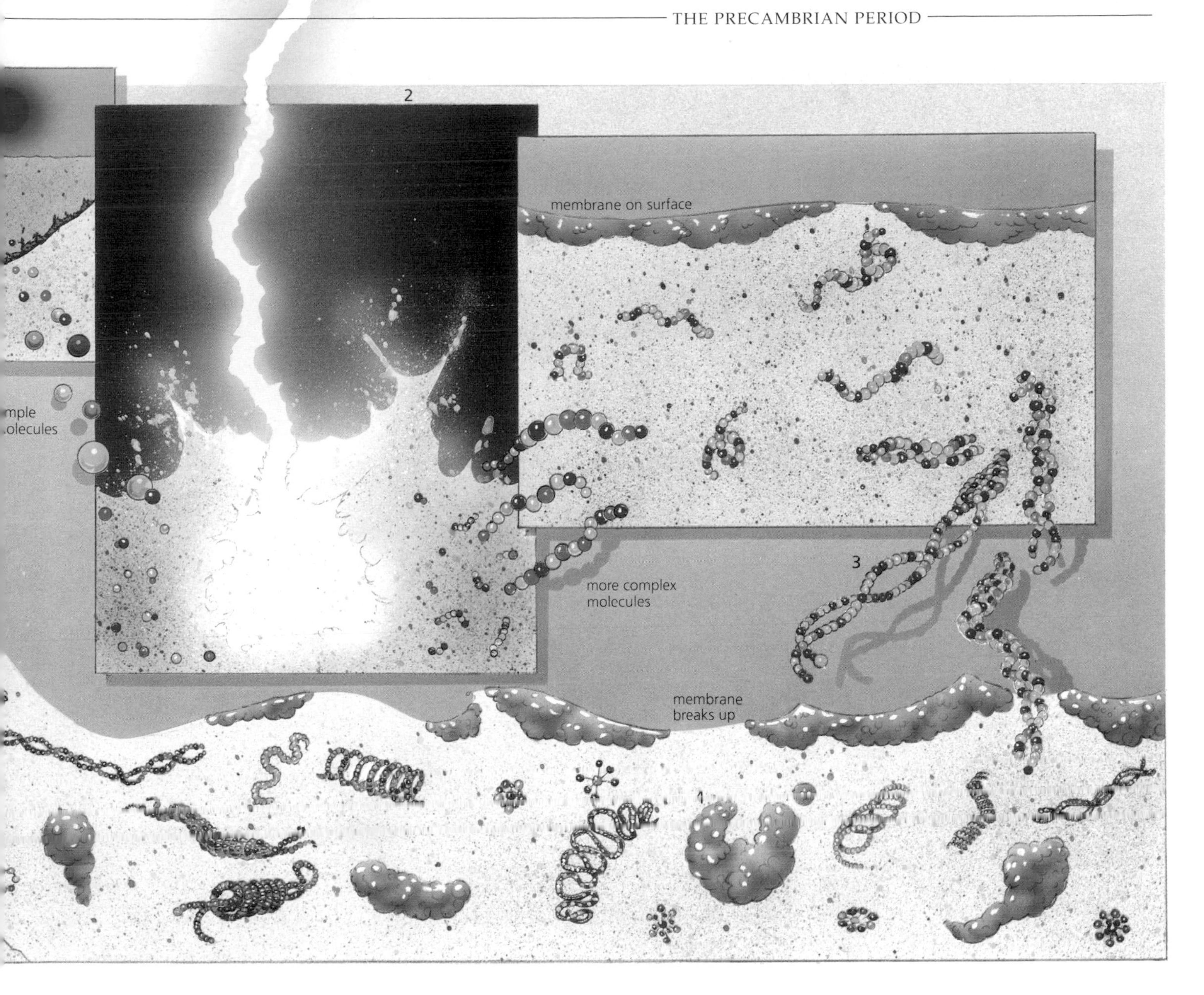

In the 1950s an American chemist, Stanley Miller, tested this theory. He mixed methane and ammonia gases over warm water and sent electrical currents through them, rather like lightning. He tried this many times, varying the mixture and the conditions. On a few occasions he found that, after just 24 hours, about half the carbon from the methane had been made into organic compounds, such as amino acids. This suggested that given enough time and the right mixture of gases, perhaps even the complex chemicals that make up DNA could be formed.

The first living cells

As the chemical soup in the early oceans became thicker, more and more new compounds were formed. Some formed membranelike sheets on the surface of the water, rather like oil forms a layer on water. When the water was then stirred, perhaps by storms, the membrane broke up into spheres, just like oil globules. Some chemicals were trapped inside the spheres, which looked rather like cells. Once DNA molecules were formed and trapped inside a membrane bag with other chemicals, life had begun.

These early cells were rather like present-day bacteria. They obtained their energy by breaking down inorganic chemicals. They could get carbon from methane, and from carbon monoxide and carbon dioxide dissolved in the water. They could get hydrogen from hydrogen sulphide and other compounds. The cells were able to use these elements to make new living material. Similar bacteria are found today around hot mineral springs and volcanoes.

Harnessing the sun's energy

The next great leap forward in the evolution of life was harnessing the sun's energy. Instead of getting energy from inorganic compounds, cells began to use the sun's rays directly.

This was the beginning of photosynthesis, the process by which plants are able to make their own food using sunlight energy. And instead of getting their hydrogen from such materials as hydrogen sulphide, they got it from something much more plentiful: water.

PHOTOSYNTHESIS – THE GREAT LEAP FORWARD

Plants, algae, and some kinds of bacteria use colored compounds called pigments in their cells to trap sunlight. They use the light energy to make all the organic compounds they need to grow and multiply. This process is called photosynthesis, which means "building up with light." When simple chemicals such as water and carbon dioxide are built up into complex compounds such as the sugars and proteins found in living cells, energy must be put in. It is rather like building a wall: You have to put in energy lifting the bricks onto the top of the wall and fixing them in place. In photosynthesis, this energy comes from light. Carbon dioxide (containing carbon and oxygen) and water (containing hydrogen and oxygen) provide carbon, oxygen, and hydrogen for the sugars, and other organic compounds that are made in photosynthesis. Not all the oxygen is used, and some is given off into the atmosphere.

▲ Primitive forms of bacteria and cyanobacteria (blue-green algae) still thrive in hot mineral-rich springs today. Some use minerals from the spring as a source of raw materials for photosynthesis. Scientists think life may first have evolved in such environments. At the bottom of the picture you can just make out two people on the walkway beside the spring.

To trap the sun's rays, these new photosynthetic cells had pigments – colored materials that can absorb light. Until now, life on Earth had been dull and colorless. Now it took on a host of new colors. Life was no longer confined to those places where special energy-providing materials were to be found, as water and sunlight were much easier to come by.

The new photosynthesizers mostly lived in mineral springs and warm water around the coasts, where the water was shallow enough to let the light through, but deep enough to protect the cells from harmful levels of ultraviolet radiation. A few still probably used hydrogen sulphide to provide hydrogen, and their descendants survive today around mineral-rich hot springs.

The age of stromatolites

Some of the earliest fossils of photosynthetic creatures are of stromatolites (see also page 34). These strange structures appear to be made up of lots of rings of limestone, with thin organic layers in between. In fact, they were made by organisms rather like very simple cyanobacteria, which are sometimes known as blue-green algae. Stromatolites come in a great range of shapes and sizes. Some are round like potatoes, while others are cone shaped, or tall and thin, or even branching.

Stromatolite fossils have been found all over the world. In many places they formed huge reefs, often several hundred feet deep in clear water, rather like tropical coral reefs today. The oldest certain stromatolite fossils have been found in rocks 2.8 billion years old in Western Australia. But structures suspected to be stromatolite fossils have been found in rocks 3.5 billion years old. Living stromatolites are still found today. They thrive in warm, shallow water, just as they did in the past. But they are limited to places where there are few modern grazing animals to eat them.

The red beds

Some of the most ancient fossils, many of them stromatolites, are found in rocks called banded-iron formations. These rocks contain layers rich in iron, unlike any later sedimentary rocks. These greatly puzzled geologists until they realized they were

caused by the activities of stromatolites. As the oxygen concentration in the oceans increased, it began to react with dissolved iron to form compounds of iron and oxygen. These oxides cannot dissolve in water, so they sank to the bottom into the sediments.

Around 2.2 billion years ago, some other new sedimentary rocks were formed on land: the red beds. These were rich in iron oxides that colored them rusty red. This shows that there was oxygen in the atmosphere by this time. The iron in the ocean had been used up and oxygen gas was escaping into the atmosphere.

Poisoned by oxygen

The oxygen in the atmosphere continued to build up during the rest of the Precambrian. But it was by no means welcome to many existing living creatures. For them, it was atmospheric pollution on a grand scale; they had evolved when there was no oxygen, and the oxygen poisoned them. Many species became extinct – the first great extinction in the history of life on Earth. It is a strange fact that, while life as we know it today cannot survive without oxygen, the first living creatures could not survive with oxygen.

Eventually, cells evolved that could not only cope with the oxygen, but actually take advantage of it. Some of the compounds made by photosynthesis can be broken down using oxygen, and the energy released can be used to make a whole new range of compounds. This is the process of respiration that takes place in most living cells today. It is called aerobic respiration (*aerobic* means "using air"). It releases a lot more energy than other biological breakdown processes that do not use oxygen. Some of the new respiring cells even devoured other cells to use them as food.

Setting the stage for evolution

As oxygen built up in the atmosphere, an ozone layer began to form, absorbing the harmful ultraviolet radiation from the sun. Life could now move closer to the water surface, and could invade the moist edges of the land. The cyanobacteria, too, were becoming more complex; they began to group together to form clumps and thin strands. But the new oxygen-respiring cells were taking over.

A new, larger, and much more complex kind of cell began to flourish about 1.2 billion years ago. Different processes were going on in different parts of the cell, contained inside separate membrane bags with special internal environments called organelles. This made the reactions inside them much more efficient. The DNA – the material containing the code of life – was organized into structures called chromosomes. Scientists think these new cells formed when oxygen-respiring cells moved into other cells, perhaps for protection from the new cell-eating cells. The new cells shared the energy and the compounds they made.

◀ A section through a fossil stromatolite, showing the layers of limestone and cyanobacteria.

◀ Living stromatolites at Shark Bay, Australia. As stromatolites carry out photosynthesis, they use up the carbon dioxide in the water. This causes calcium carbonate (lime) to come out of solution. Tiny particles of lime are trapped by sticky mucus produced by the stromatolites, forming layers of limestone.

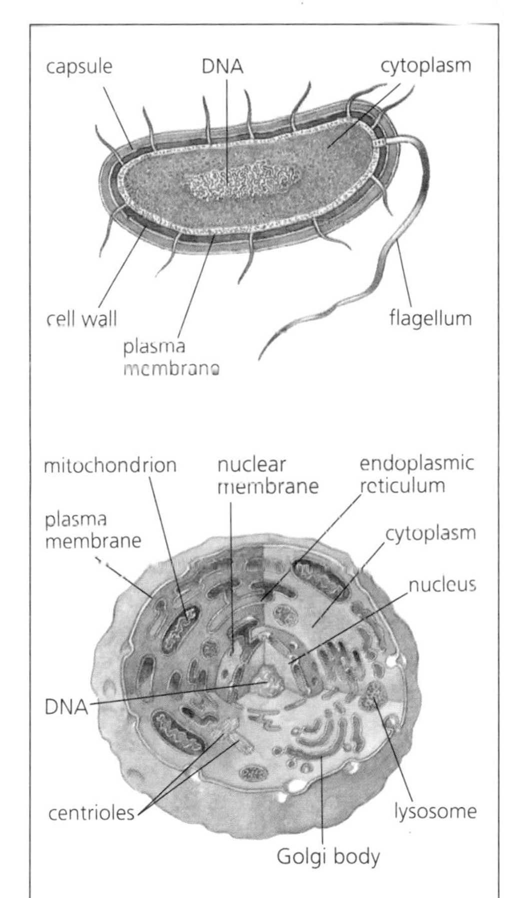

▲ The earliest cells, called procaryotic cells (above), were very simple. All the cell chemicals, including the genetic code in the DNA, were mixed up in the cell. In later cells, called eucaryotic cells (below), little membrane bags inside the cells housed the chemicals for particular reactions, providing just the right environment for these reactions to go at their fastest rate. The DNA was arranged on chromosomes inside a nucleus surrounded by a membrane. The nucleus controlled all the activities of the cell.

Variety, the spice of life

More important still, the cells began to reproduce in a new way. Instead of simply dividing in two, and producing two cells exactly like their parents, the new cells did something very strange. Two cells joined together, swapped some of their DNA, then divided to form two or more new cells. This is sexual reproduction. The new cells contained a mixture of DNA from both their parents. Sexual reproduction produced more variation among the cells, which greatly sped up evolution.

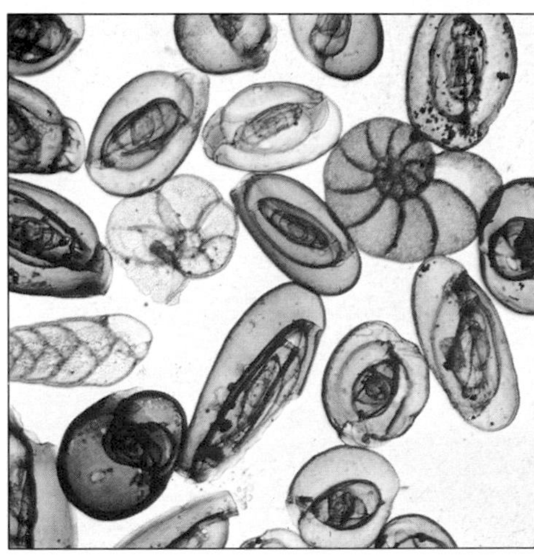

▲ A great variety of single-celled creatures live in the surface waters of the oceans today. Many are probably very similar to forms that existed in Precambrian times.
Top: These are the microscopic glassy skeletons of radiolarians, single-celled animals with long fine spines covered in sticky mucus to trap tiny prey animals.
Bottom: The chalky, many-chambered shells of foraminifers are useful as index fossils. The shells make up the bulk of certain types of limestone. Like radiolarians, the single-celled foraminifers have long sticky spines for trapping prey.

The first great extinction

The late Precambrian was a time of great upheavals. There were many volcanic eruptions, earthquakes, and bouts of mountain building. The huge quantities of volcanic ash that entered the atmosphere caused a cooling of the climate, and as the land masses moved over the pole, great ice caps spread across the globe.

Many species became extinct during this period. Finally the ice began to melt, and as it melted the sea rose and flooded the edges of the continents. For the shallow-water creatures a host of new unoccupied habitats opened up, with new opportunities for specialized life-styles. By now far less harmful ultraviolet radiation from the sun was reaching the earth's surface, because it could not penetrate the thickening ozone layer. There was also more oxygen in the atmosphere, which suited the newly evolving animals.

The mystery of the multicells

Nobody knows just how the first of the many-celled (multicellular) animals arose. Perhaps the cells divided but did not quite separate. Or perhaps individual cells came together and organized themselves. This is not so farfetched as it sounds. In 1907 a biologist named H. G. Wilson did some experiments with sponges. He cut up a red sponge into tiny pieces, then forced the pieces through a muslin bag to separate the cells until he had just a red sediment in a jar of water. To his surprise, within hours the cells had clumped together. As the days passed, they began to organize themselves into a new sponge, forming chambers and canals and branching tubes. Within a week the sponge was as good as new. Perhaps this is how multicellular animal life began.

There are also strange creatures called slime molds, which look like colored blobs of slime that creep over the ground, or over the bark of trees. One group of slime molds, the cellular slime molds, spend most of their lives as independent cells that creep around in the soil, where they feed on bacteria. But when food supplies run short, they produce a chemical that attracts other slime mold cells. Millions of cells come together to form a great mass of cells rather like a multicellular animal. This moves around and responds to light and chemicals as if it were a single animal. Eventually it forms itself into a fruiting body rather like the spore cases of some fungi. There is a tall stalk with a protective outer coating, and a bag of spores at the top.

Marks in the mud

These early soft-bodied animals were not easily preserved. But they have left their footprints, or rather, their trails, in the rocks. Feeding burrows, surface trails, and resting marks in the mud have been found in rocks

One creature or many? In response to a chemical signal, millions of amoeba-like slime mold individuals come together to form a moving sheet, which eventually puts out stalked spore capsules rather like those of simple fungi.

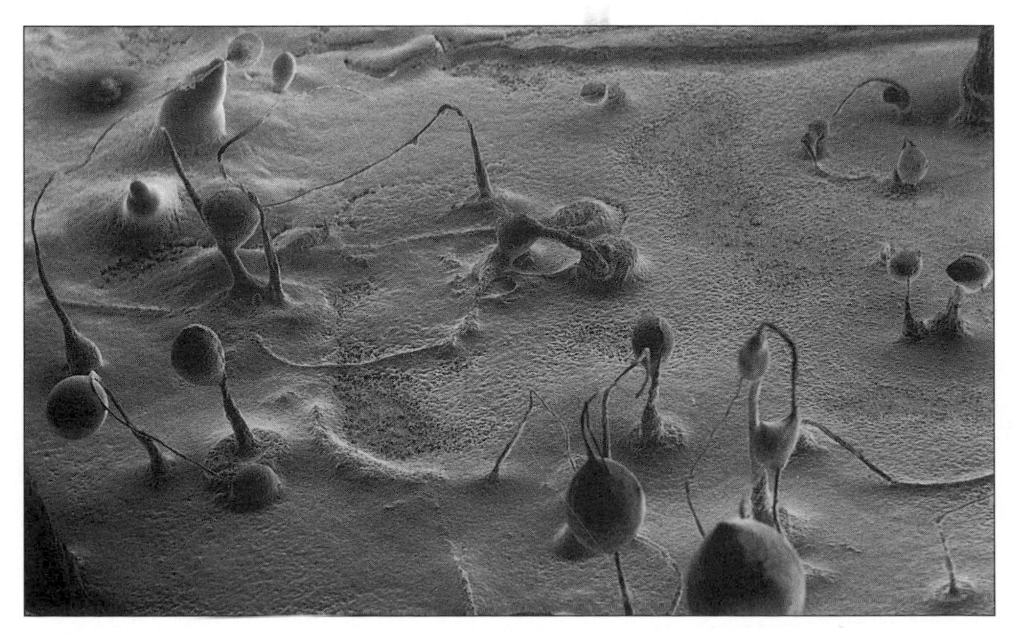

dating back 700 million years or more. But very few such clues were left until 640 million years ago, when the late Precambrian ice age ended and the scene was set for a great burst of evolution.

The Ediacara animals

In a remote part of southern Australia, in the Ediacara Hills, are some ancient shallow-water and seashore sediments that have been dated to 640 million years ago. Here a remarkable number of Precambrian animals have been preserved. There are at least 30 genera of multicellular animals in these rocks, and similar communities of fossils have been found in rocks of the same age in many parts of the world.

The Ediacara animals lived mainly on the seabed. They fed on a layer of organic material (detritus) covering the mud, the remains of a multitude of single-celled creatures living in the surface waters above. Flatworms and segmented worms swam just above the surface or crawled across the sediment. There was no need for great speed, since there were few predators (animals that feed on other animals).

Sea pens rose like feathery flowers from the seabed to filter the water. Tube worms lay in the sediment, extending their tentacles to sweep the detritus-rich water. Some primitive echinoderms, relatives of modern starfish and sea urchins, lived in the mud. There were also many large, flat, pancake-shaped animals, rather like jellyfish, that apparently lived on the mud. Above them, true jellyfish drifted in the open sea.

Clues to the future

The Ediacara deposits contain many trace fossils, evidence of the churning of the seabed by soft-bodied animals. In places there are paired V-shaped markings in the mud, like the scratch marks left by pairs of tiny legs. These may have been the tracks of primitive arthropods or joint-legged animals, the group from which the fossil trilobites and our modern insects, spiders, and scorpions are descended. However, there were no hard remains, so they had probably not yet evolved hard shells.

▼ The Ediacara animals were all soft bodied. There were many different kinds of jellyfish (1). *Dicksonia* (2) and *Spriggina* (3) were flattened, wormlike animals. *Spriggina* had many little paddles along its sides for swimming, like marine worms today. It may have been an ancestor of the trilobites. *Charniodiscus* (4), *Rangea* (5), and *Pteridinium* (6) were leaflike sea pens, colonies of tiny hydralike animals that filtered food particles from the water. *Tribrachidium* (7) is a mystery; it has a Y-shaped central mouth and bristlelike appendages on its surface, and may have been an ancestor of the echinoderms.

THE CAMBRIAN PERIOD

570 MILLION TO 500 MILLION YEARS AGO

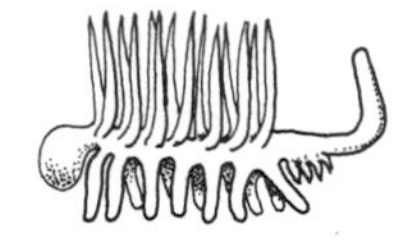

The Cambrian period began about 570 million years ago, or maybe earlier, and lasted for 70 million years. The period began with an astonishing explosion of evolution, during which most of the major groups of animals we know today made their first appearances on Earth. The boundary between the Precambrian and the Cambrian is the point in the fossil record where a great variety of animals with mineral skeletons suddenly appears – "the Cambrian explosion" of life.

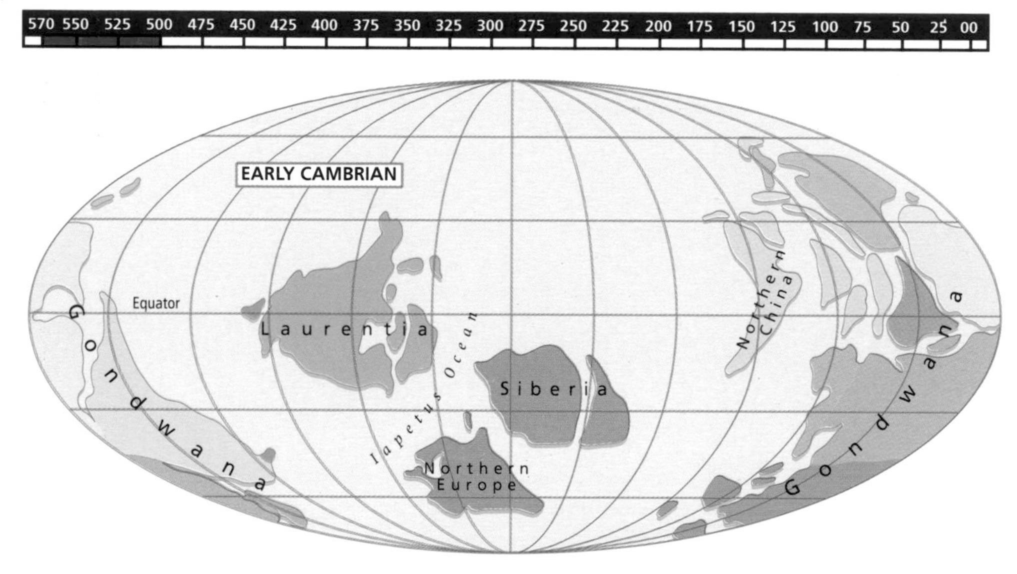

We do not know exactly what the map of the world looked like in Cambrian times, but it was very different from the world we know today. Lying across the equator was a huge continent, Gondwanaland, made up of parts of present-day Africa, South America, southern Europe, the Middle East, India, Australia, and Antarctica. There were also four smaller continents equivalent to present-day Europe, Siberia, China, and North America (but with northwest Britain, western Norway, and parts of Siberia tacked on). The North American continent is known as Laurentia in this period.

The climate was warmer than today. Around tropical shores were great reefs of stromatolites, much as coral reefs grow in shallow tropical seas today. But the reefs were already shrinking, as the newly evolved multicellular animals began to eat them. The land was bare of plants and soil, so erosion by water and wind was much faster. Large amounts of sediments were washed into the sea.

The riddle of the skeletons

Until animals developed hard skeletons, they were not easily preserved as fossils, so there are few records of them. But why did so many evolve skeletons now, and not earlier? It seems that an animal needs a certain amount of oxygen in order to lay down the minerals needed to form a skeleton. Perhaps the early Cambrian was the first time when there was enough oxygen in the atmosphere.

The earliest skeletons were mainly of calcium carbonate. As new predators began to graze on the old stromatolite reefs, the reefs shrunk, releasing more calcium into the oceans, which could be used to form skeletons and shells. As well as providing support, the shells gave the animals protection against the newly evolving predators.

Stiffer skeletons also opened up new ways of life: They helped the animals lift themselves above the mud to move faster over the seabed. Once jointed legs had evolved, all sorts of new ways of life were possible, such as walking and swimming. Bristly legs could be used for filter feeding, and jointed mouthparts opened up new ways of tackling prey.

The Cambrian explosion

The Cambrian explosion is one of the biggest mysteries in the history of life on Earth. It took 2.5 billion years for simple cells to evolve into the more complex eucaryotic cells, and another 700 million years for multicellular creatures to appear. Then, in just 100 million years came a tremendous variety of multicellular animals. Since then, for over 500 million years, not a single new phylum (basic body plan) has evolved.

During the Cambrian there was a great area of shallow continental shelf covered by soft muds and warm water – ideal conditions for life. There was by now plenty of oxygen in the atmosphere, though not as much as there is today. The evolution of hard coverings made new life forms possible, such as the arthropods. Animals needed new ways of protecting themselves against the new predators. And as their defenses improved, the predators had to evolve new methods of hunting to overcome them.

Throughout the Cambrian the sea level rose and fell, making some populations extinct, then providing new habitats for other animals to colonize and adapt to. As the Cambrian period went on, animals evolved new, more specialized ways of feeding. This led to a greater variety of animals, and meant that more could live side by side without competing for the same food sources. Never again would there be so many empty habitats and so little competition – so many opportunities for nature to experiment.

The Burgess shale

In 1909 an American paleontologist, Charles Doolittle Walcott, made one of the discoveries of the century. About 7,900 miles up in the Canadian Rockies he found a narrow bed of shale that contained an extraordinary number of very strange fossils of soft-bodied animals, many of them remarkably well preserved. These creatures had lived on shallow-water mud banks at the side of an early Cambrian reef. They were swept into deep-water muds and buried rapidly by a mud slide, also trapping on the way some of the animals that lived in the water above the reef.

Scientists think the Burgess shale was formed at the very start of the Cambrian period. It contains all sorts of animals not found in older rocks. There were arthropods that crawled over the mud, feeding on detritus (organic debris), and others that were actively swimming filter feeders. Some swimming arthropods, such as *Sidneyia*, were probably predators. Other animals lived on and in the muds. These included many different kinds of sponges, some with long spikes to which brachiopods (lampshells) clung to filter the water. Soft-bodied sea pens and armor-plated sea lilies waved in the currents, and primitive segmented worms swam by beating fringes of bristles.

▶ Animals of the Burgess shale. *Eldonia* (1) jellyfish drift among treelike glass sponges (*Vauxia*) (2). The strange arthropods *Protocaris* (3) and *Plenocaris* (4) swim past a *Mackenzia* (5), thought to be a kind of sea anemone. It is dwarfed by the huge predator *Anomalocaris* (6), whose powerful mouth was probably capable of crushing the shells of other arthropods. Crustaceans like *Burgessia* (7) and *Canadaspis* (8) feed on the mud, sifting out food particles. *Naroia* (9) was a primitive, soft-bodied trilobite, while the weird *Wiwaxia* (10) was actually a kind of annelid worm, covered with plates and spines, and so was *Canadia* (11). Stranger still are *Opabinia* (12) and *Hallucigenia* (13), unlike any animals alive today, and the wormlike *Odontogriphus* (14), which actually had a horse-shaped mouth surrounded by tiny teeth and tentacles.

A bizarre collection of beasts

Walcott identified some 70 genera and 130 species of animals in the Burgess shale. He used local Native American words to name many of the animals: *Wiwaxia*, for example, was the word for "windy," a good description of the site, while *Odaraia* comes from *odaray*, which means "cone-shaped." The animals are as bizarre as their names. Some can be placed in modern groups, but others look nothing like any other known creature, past or present.

Hallucigenia was an odd creature, with a bulbous "head" and a row of spines along its back. *Opabinia* had five eyes, four of them on stalks, and a long flexible nozzle that may have sucked up detritus from the seabed. The tip of the nozzle forked in two and was armed with spines. Was it used like pincers for grabbing food? Or did it simply pass food back to the mouth?

Some animals seem to have features of more than one modern phylum. *Odontogriphus* looked like a flattened segmented worm, but had arthropodlike feelers beside its mouth, and many tiny teeth. *Nectocaris* had a head and upper body like a shellfish, but a lower body and tail like a backboned animal.

The great experiment?

It is almost as if nature were experimenting with many different animal forms during the "explosion of life" in the Cambrian period. However, only a few of these forms have survived to the present day. There were many strange phyla and body plans in the Cambrian that no longer exist today. There were also many familiar groups of animals. In fact, by the end of the Cambrian, all the modern hard-bodied animal phyla except one had evolved.

But why have no more phyla evolved? Have the genetic systems of animals altered, so that they cannot change so easily? Or is there so much competition from the many existing species that there is little opportunity for experimentation?

A scene from the late Cambrian shallow-water seabed. The trilobites (*Paradoxides* (1), *Bailiella* (2), *Solenopleura* (3), *Hyolithes* (4), and *Agnostus* (5)) are very much in evidence. Sea pens (6), archaeocyathids (7), and drifting graptolites (8) (*Dictyonema*) comb the water for food, while brachiopod mollusks (*Lingulella* (9) and *Billingsella* (10)) draw water in through their shells to filter it.

Certainly, any new living spaces that appear are quickly invaded by existing animals already suited to live in them.

Life in Cambrian seas

The early Cambrian burst of evolution produced many different creatures. The most important animals were the trilobites, joint-legged animals rather like modern horseshoe crabs, with shieldlike shells on their backs. Most of the early trilobites lived on the seabed, but a few swam in the waters above and may well have preyed on the mud dwellers.

There were also many other creatures in the water. Millions of floating algae and microscopic animals formed the basis of the food chain: the series of living things each dependent on the next for food. A few of these, such as the foraminifers and the seed shrimps that had arisen in the Precambrian, were evolving hard shells. Jellyfish and their relatives pulsed through the sea, and by the end of the Cambrian there were advanced predators such as cephalopod mollusks (like modern-day octopuses and squid) and primitive armored fish.

Burrowing and scavenging worms worked the mud, together with primitive mollusks, rather like modern limpets and sea snails, and brachiopods. Brachiopods are filter-feeding animals with two hinged shells rather like cockles on stalks. Forests of sea pens filtered the water above the seabed, and delicate glass sponges survived in still water. By the end of the period many different echinoderms had appeared, including starfish and sea urchins.

All change on the reefs

Predators were busy destroying the old Precambrian stromatolite reefs, but new limestone producers were at work. These were the archaeocyathids, simple spongelike animals that soon spread and evolved into many different species. These suddenly declined and became extinct in the middle of the period, but by this time the first corals had appeared, although they were not yet building reefs.

At the end of the Cambrian period there was a new ice age, and the sea level fell, destroying many habitats and causing many species to become extinct.

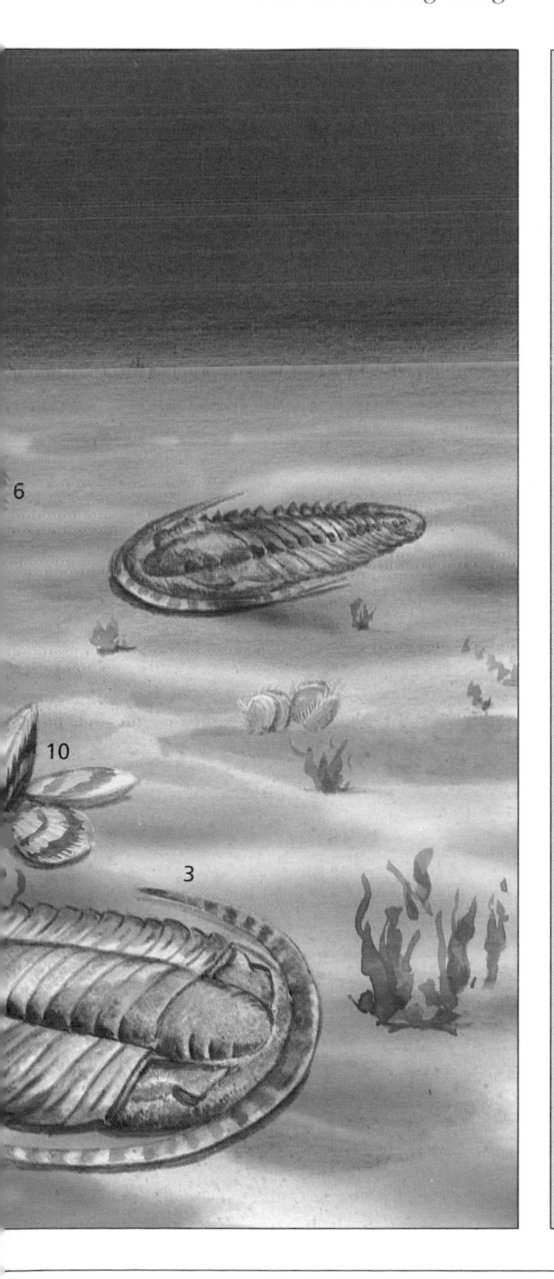

A HUMAN TAIL

The Cambrian saw the appearance of the chordates, the group of animals that eventually gave rise to human beings. Chordates have gill slits at some stage in their lives and a distinctive main nerve cord running down their back, with paired blocks of muscles on either side of it. Later the nerve cord became surrounded by bony vertebrae to form the backbone of the vertebrates. This backbone extended beyond the anus as a tail. Chordates also have a stiff rod of cartilage (the notochord) running down the animal's back at some stage in its life cycle. The notochord is still present in the embryos of vertebrates, including humans.

There are three possible early chordate groups in the Cambrian. All were somewhat fishlike in shape, and the nerve cord reached into a long tail powered by V-shaped blocks of muscles. Behind the head were gill slits. Similar animals survive today: the tadpolelike larvae of sea squirts and the adult lancelet.

The first potential ancestor of the chordates was a small fishlike animal called *Pikaia* from the Burgess shale. It looked rather like a lancelet, with a long bar running down it and segments that looked like muscle blocks. Later came the calcichordates, which had a chalky outer skeleton with a distinct head and tail, and the conodonts, which had a tail fin with V-shaped muscle blocks, and a structure rather like the mouth of a jawless fish, with teeth made from enamel and dentine like those of the vertebrates. By the end of the period the first vertebrates, the pteraspid fish, had appeared.

▲ Two living lancelets

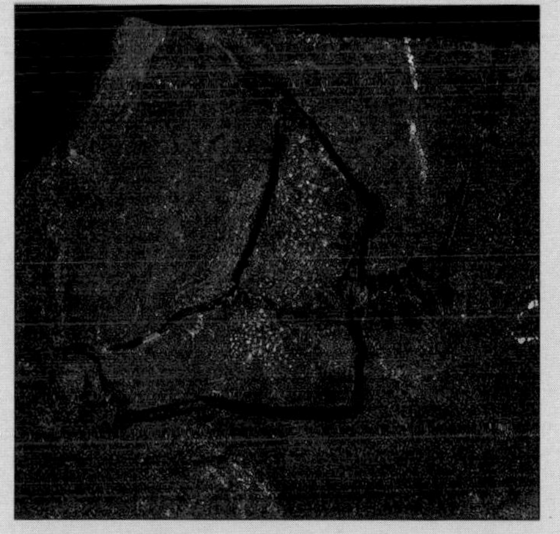
▲ A fossil calcichordate

Terrific Trilobites

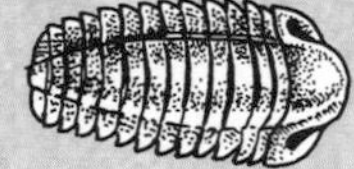

Trilobites dominated the Cambrian seas. They burrowed in the sediments, crawled over the seabed, paddled through the murky bottom water, and swam in the sunlit upper waters. Many were scavengers, feeding on dead animals and detritus in the sediments, but others were predators. Some may even have preyed on their mud-dwelling relatives. The largest trilobites were over 27 inches long, the smallest less than half an inch.

Trilobites looked rather like modern king crabs (horseshoe crabs), to which they are distantly related. The name *trilobite* means "three-lobed." This is because the shell was in three sections – a central ridge with a flattened area on either side of it. Most trilobites had a shieldlike head, a flexible hinged thorax (middle section), and a flattened tail, which was sometimes drawn out into long spines. Some fossil trilobites appear to be rolled up like wood lice, perhaps for extra protection against their enemies.

On each segment of the body was a pair of limbs. The pair in front of the mouth were used as feelers. The other limbs had a feathery gill for breathing, a swimming paddle or a walking leg, and a spine that was used to pass food along the body to the mouth. The shell was often shaped into ridges and knobs. Some trilobite shells are covered in tiny holes, perhaps where there were hairs for feeling or tasting.

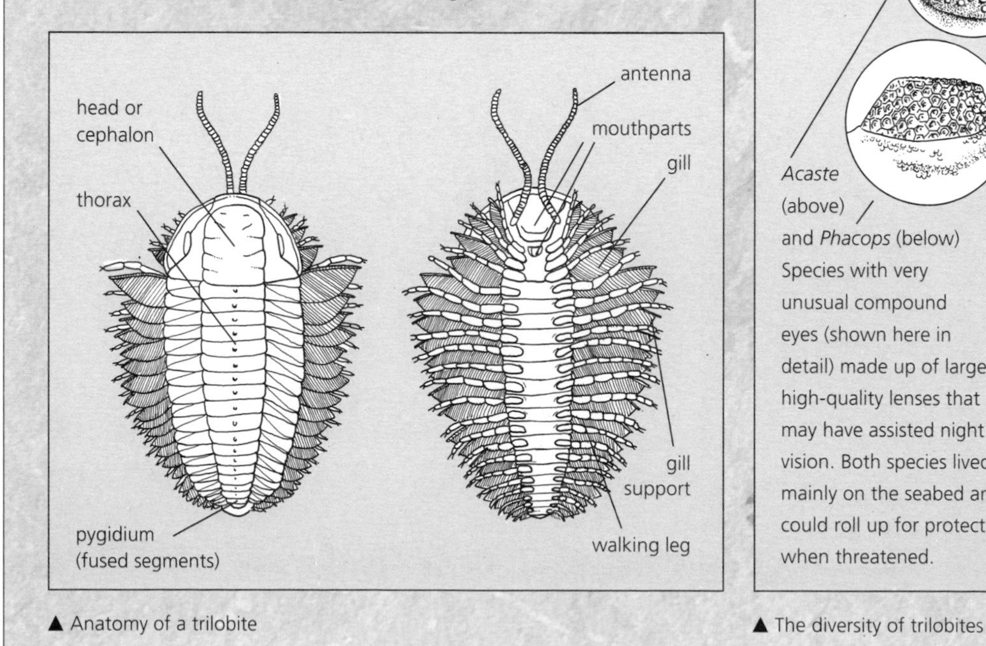

▲ Anatomy of a trilobite

Olenellus An early Cambrian trilobite with many spiny body segments. These would have had gills that may have helped it breathe in oxygen-poor water or mud.

Agnostus A very small trilobite, less than half an inch, with no eyes. It probably lived in the seabed sediments, or in deep water where there was no light.

Miraspis Used its spines to prop itself up above the sediment, probably feeding on small creatures in the water.

Cryptolithus A blind species. Its strong legs could excavate shallow burrows, in which the trilobite sifted food from the water currents.

Bumastus A stiff animal with limited powers of movement. It probably lived in sediments, filtering food from the water.

Deiphon Probably a plankton feeder, living in the surface waters of the sea. Its long spines may have helped it float.

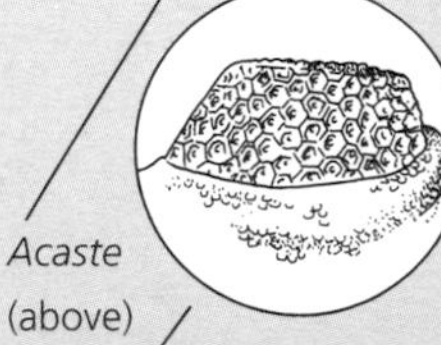

Acaste (above) and *Phacops* (below) Species with very unusual compound eyes (shown here in detail) made up of large, high-quality lenses that may have assisted night vision. Both species lived mainly on the seabed and could roll up for protection when threatened.

Harpes The broad, flat rim of the cephalon may have acted rather like a snowshoe to spread the trilobite's weight on the soft sediments.

▲ The diversity of trilobites

Pieces of the past

Like all arthropods, trilobites had a hard outer covering that they had to shed (molt) in order to grow. Their cast-off skeletons made good fossils. But to help them molt, the trilobites had lines of weakness, called sutures, in their shells. When they were buried in the sediments, the shells tended to break up along these lines, so it is rare to find whole skeletons.

Trilobite detective work

How do we know how trilobites lived? The remains of their mouthparts and front legs provide clues to their feeding habits. But did they swallow the sediment and strain out food from it, or did they feed directly on the detritus at the surface? And how did they move? Did predatory trilobites chase their prey or lie in wait for it?

Some of the answers are to be found in trace fossils – the tracks the trilobites left as they moved around. When they plowed through the mud, they left a herringbone pattern. When they rested, they made a track like a hoofprint.

The first eyes on Earth?

The trilobites were the first fossil animals to have well-developed eyes. They were probably used to help the trilobites spot predators. Like modern insect and crustacean eyes, these were compound – made up of clusters of tiny lenses. These were hard enough to survive as fossils.

Trilobite eyes come in many different sizes and shapes. Some trilobites were completely blind, perhaps because they lived in the sediments or in deep water where there was little light. Others had wraparound eyes with a wide field of view. In some the eyes were at the sides of the head, but in others they bulged near the top of the head or stuck up on stalks, perhaps so they could stay almost hidden in the sediments but still keep an eye out for danger, or for prey. Actively moving trilobites had bulging eyes at the front of the head. The two eyes would have had overlapping fields of vision, allowing the trilobite to judge distance and speed.

Swimming trilobites evolved large, flattened tail plates. These species had light shells and many spines to increase their surface area and help them float more easily. Deep-water species used spines to prop themselves up above the sediment, perhaps in order to filter food from the water.

Demise of the trilobites

The trilobites reached their greatest development in the Ordovician period, but by the end of the Paleozoic era, 225 million years ago, they were extinct. The rapidly evolving mollusks and fish were able to attack them despite their shells, and also competed with them for food.

▲ The trace fossil *Cruziana* was actually made by a crawling trilobite.

SEALED IN ARMOR

Some trilobites were able to fold up, so that their heavy armor plating completely protected their more vulnerable underside.

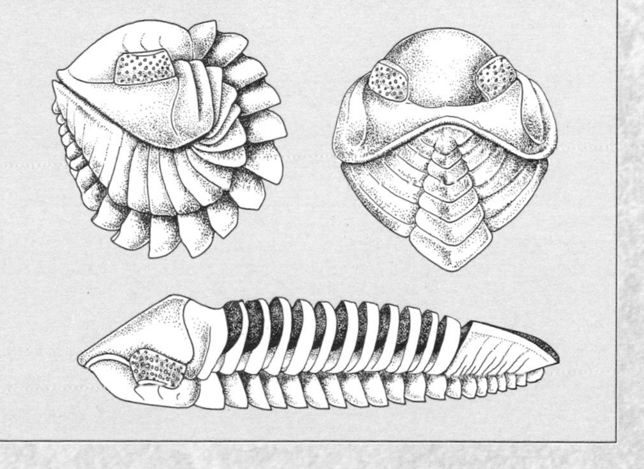

THE ORDOVICIAN AND SILURIAN PERIODS

500 MILLION TO 438 MILLION YEARS AGO

At the start of the Ordovician period, there was still the great continent of Gondwanaland in the Southern Hemisphere, while the other large landmasses lay close to the equator. Europe and North America (Laurentia) were being pushed apart by the expanding Iapetus Ocean. This ocean reached a width of about 1,240 miles, then began to shrink again, as the land masses of Europe, North America, and Greenland drifted closer together until they collided. During the Silurian period, Siberia moved to join Europe, Africa collided with southern North America, and the giant supercontinent of Laurasia was born.

570 550 525 500 475 450 425 400 375 350 325 300 275 250 225 200 175 150 125 100 75 50 25 00

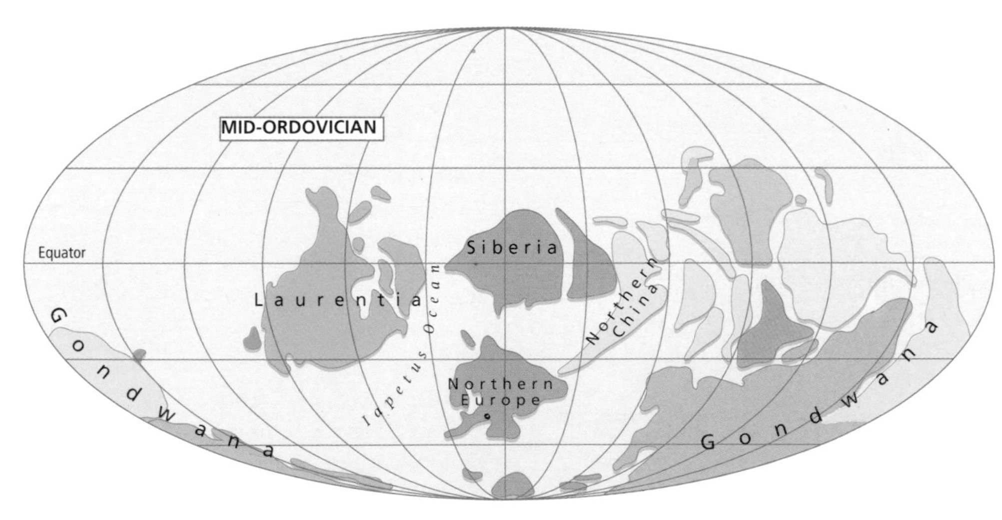

The early Ordovician period was a time of rising sea level, as the old Cambrian ice caps melted. With so many of the world's landmasses in the tropics and subtropics, shallow seabed animals and reefs flourished around the edges of the continents.

As the Ordovician period gave way to the Silurian period, a new ice age arrived. The Silurian period lasted from 438 to 408 million years ago. When the ice finally melted, the sea level rose to flood many parts of the continents and the climate became milder. Later the water retreated again, due to movements of the earth's plates.

These were not the only changes going on. The movements of the continents were accompanied by many volcanic eruptions and earthquakes, and great ranges of mountains were pushed up. The remains of these mountains survive today as the Urals in Russia, the mountains of Norway and Scotland, and the Appalachians along the eastern part of North America. These changes in the pattern of land, sea, and climate led to many species becoming extinct.

Lilies of the sea

The Ordovician seas had very different occupants from the ancient Cambrian seas. The evolution of hard shells meant that animals could now lift themselves above the sediments to reach the food-rich waters above. During the Ordovician and Silurian periods, more and more filter-feeding animals evolved. Among the most attractive were the crinoids (sea lilies), which were rather like armor-plated starfish on stalks, swaying in the currents. Crinoids have long flexible arms covered in sticky tube feet that trap food particles from the water. Some had as many as 200 arms.

Crinoids and their stalkless relatives, the feather stars, survive to this day.

Land of the lampshells

An Ordovician beachcomber would have found the beach dotted with brachiopods, some of the most successful filter feeders of Ordovician and Silurian times. Some species were shaped like Roman lamps. These had a scoop-shaped base filled with oil, covered with a curved lid, hinged at one end. The oil, when burning, provided light. Brachiopods have two shells joined by a hinge, and look rather like fat clams.

Inside the body of a brachiopod is a long spiral structure fringed with tentacles coated in microscopic beating hairs called cilia. This is the lophophore, whose function is to waft water into the animal and filter out food particles and oxygen. Many brachiopods were attached by a stalk or even directly by their shells to the seabed, while others simply lay on the sediment.

▲ A small corner of the Ordovician seabed. Nautiloids (1) hunt among the waving crinoids (sea lilies) (2). *Echinosphaerites* (3) is another kind of stalked echinoderm, while *Bothriocidaris* (4) is more like a sea urchin. This is the age of trilobites – a great variety of them (*Brogniartella* (5), *Tetraspis* (6), and *Platylichas* (7)), sift the sediments for food. The mollusks are diversifying, too. *Lophospira* (8) and the bellerophontids (9) feed on detritus or the occasional corpses. The filter-feeding brachiopods (*Platystrophia* (10), *Onniella* (11), and *Stophomena* (12)) use a muscular foot to anchor themselves in the sediments; while *Christiana* (13) simply rests on the convex half of its shell. The recently evolved mollusks, such as *Modiolopsis* (14), use tough byssus, or threads, to attach themselves to the rocks.

◀ A "living fossil" – a modern living brachiopod. *Lingula* is a survivor of the most primitive group of brachiopods. Its shells are made of phosphate rather than carbonate. *Lingula* lives in vertical burrows near the tide line.

◀ Fossils of curved (*Rastrites*) and straight (*Climacograptus*) graptolites. Graptolites were colonies of tiny hydralike animals (inset) that sifted food particles from the water with rings of tiny tentacles. Some graptolites were anchored to the seabed, while others floated upside down from seaweeds or simply drifted in the water.

Rise of the graptolites

In the surface waters of the oceans another group of filter feeders was flourishing. These produced strange sticklike fossils, often V-shaped or spiral, edged with rows of "teeth." In fact, these were not teeth at all, but tiny cups containing colonial animals that probably put out short feathery tentacles to filter the water. Some paleontologists even think they may have been related to the early chordates. Graptolites first appeared in the Cambrian, but became much more common in the Ordovician before they became almost extinct at the end of the Silurian period. Most graptolites were attached to the seabed, but a few species hung down from floating clumps of seaweed, and some floated free, fishing for microscopic floating animals and algae.

The new reef builders

The old reef-building archaeocyathids had died out by the start of the Ordovician. Only the ancient stromatolites remained, building small mounds on the ocean floor. But the reefs were about to get a new lease of life. Some new reef builders were emerging. The stromatoporoids were a curious group of animals that may have been sponges or relatives of the corals. They certainly looked like sponges, with many tiny holes or pores over their surface through which water currents passed. Their lime skeletons came in all shapes and sizes, from huge round lumps to narrow columns and slender branching forms.

The corals also appeared at this time. These creatures are animals called polyps, rather like sea anemones. They build lime skeletons to support their bodies. In the middle Ordovician came the rugose corals. These were the real reef builders. Single rugose corals had horn-shaped skeletons with many hard internal ribs to support them. This made their skeletons very strong – strong enough to form reefs. Colonies of rugose corals built huge reefs, especially in the Devonian period.

Builders' helpers

A strange group of colonial animals also arose at this time, and helped the stromatoporoids and corals to build reefs.

These were the bryozoans (sea mats), also called ectoprocts. Modern bryozoans form matlike coatings on rocks, seaweeds, and other objects. Through hundreds of tiny tubes in the mat, often barely 1/25 of an inch across, little animals put out rings of tentacles to sweep food into their mouths. Some of the ancient bryozoans formed thick crusts and huge dome-shaped mounds on the seabed. Others were more slender and branching, but their crumbling remains helped to fill in cracks in the reef and bind it together.

Powerful predators

In the water above the reefs, some powerful new predators were on the move. These were cephalopod mollusks called nautiloids, the ancestors of modern squid and octopus. They moved by jet propulsion. Some were probably scavengers on dead animal remains, but many were predators. Unlike the few surviving modern nautiloids, these ancient nautiloids mostly had straight or only slightly curved shells. Some grew up to 30 feet long – the largest shells ever produced by invertebrates. Gas-filled

▼ Modern coral reefs are built by corals of a very different kind, which come in many different shapes and sizes.

NAUTILUS, A LIVING FOSSIL

▼▲ These pictures show a living nautilus (below) and its fossil ancestor (above).

One genus of the nautiloids still survives today: *Nautilus*. There are several species in the west Pacific Ocean. *Nautilus* has a coiled shell divided into chambers. It grows by adding new chambers to the shell. Older chambers are filled with gas, but young ones contain liquid. To change depth, *Nautilus* changes its density by altering the amount of liquid in the chambers.

Nautilus has many features in common with its relatives, the octopus and squid. It swims backward by jet propulsion, a good trick for getting away from predators. Below its head is a ring of tentacles armed with suckers for seizing prey. Behind the tentacles is a muscular tube that can shoot out water, propelling the animal backward. *Nautilus* has a well-developed brain and good senses of taste and touch. It has large, but rather simple, eyes on stalks. *Nautilus* comes out at night to hunt for fish and shellfish, and sinks to the seabed to rest by day.

chambers helped them remain afloat. The nautiloids had hard beaks capable of crushing the shells of trilobites and their relatives. Around this time the trilobites began to develop heavier, tougher armor, perhaps for protection against the nautiloids.

The narrow, pointed shells of the nautiloids were often decorated with intricate patterns of ridges and grooves, and may also have been colored. The shells were divided into chambers separated by supporting plates. Grooves on the outside of the shell show where the plates were attached. These grooves show up as patterns on the shells of fossil nautiloids.

The nautiloids were a very successful group. Many different kinds of nautiloids arose during the Ordovician period, and in Silurian times they began to develop curved and even coiled shells.

◀ Early jawless fish, such as these *Astraspis*, had no fins to stabilize them in the water, and probably swam rather like tadpoles. The lower body was covered in rows of knobbly, bony scales, but the head end was protected by a heavy shield, probably to protect against attacks by giant sea scorpions. With no jaws or teeth, these fish simply scooped or sucked up food particles from the mud.

◀ Acanthodian fish. Small but fierce, the acanthodians were the first fish with jaws and teeth. Instead of body armor they were covered with small overlapping scales rather like those of a modern fish.

Fish in armor

The Ordovician and Silurian seas had very different occupants from those of the old Cambrian seas. The primitive fish that had first appeared in the late Cambrian were multiplying and changing. They had no jaws, and simply sucked up detritus from the mud. Or perhaps, like the modern jawless lampreys and hagfish, some may have fed on carrion (dead flesh) or lived as parasites on other fish. The large, bony scales of these early jawless fish later evolved into teeth and jaws.

These early vertebrates did not have bony internal skeletons like those of most modern vertebrates. But many were partly or wholly covered with a massive armor of bony plates, especially at the head end. They have been described as "like a crab in front and a mermaid behind." They were called ostracoderms, which means "bony shield." The tail was coated with smaller plates and was much more flexible. Its powerful movements provided propulsion for swimming. The bony plates formed a kind of external skeleton. In modern bony fish – and in humans – the skeleton is first formed as cartilage, which is later substituted by bone. Cartilage is a softer, more flexible material than bone. It does not preserve well, so we do not know if these early fish had a skeleton of cartilage.

By Silurian times, fish looked more fishlike. Most of them now had fins, and instead of bony armor they had small scales. By the early Devonian, or perhaps earlier, several different groups of bony fish had evolved.

Fearsome fish

In the early Silurian a group of small fish, the acanthodians, became the first fish predators. The name *acanthodian* means "spiny" – the fins were supported by stiff spines, perhaps used to make it difficult for predators to swallow them.

The acanthodians are the first fish with jaws to appear in the fossil record. An acanthodian's mouth resembled a dentist's nightmare! Their jaws could gape wide, but their throats and the stiff arches supporting their gills were armed with spikes, presumably to grip prey being swallowed, or perhaps to help in filter-feeding. Most acanthodians had teeth, perhaps for seizing prey. Toothless species were probably filter feeders. The acanthodians were covered in small, thin, overlapping scales similar to those of many modern fish.

The first land plants

The land was still a hostile place to live, a barren landscape of roaring volcanoes, dry rocky plains, strong winds, and fierce sunshine. There was no soil and no shade, for there were no plants.

The first of the true plants appeared during the late Ordovician period. For millions of years branching red, green, and brown algae (seaweeds) had flourished in the shallow coastal seas. Communities of algae, fungi, and bacteria crept out on to the mud, and began breaking it down to form a primitive kind of soil, which paved the way for plants to move in. Perhaps the first plants evolved along the shores of lakes or drying swamps. In danger of being stranded by falling water levels, some green algae developed a waxy surface (called a cuticle) to prevent them from drying out when the water was low. Small openings in the cuticle, called stomata, let the carbon dioxide in for photosynthesis and the oxygen out.

Surviving out of water

As the plants moved out of the water, they needed a new source of water and minerals. The threads that held them in the sediment evolved into real root structures that could absorb water and minerals from the mud. A network of tiny tubes (xylem) carried water from roots to stem, and another set of tubes (phloem) carried the materials made by photosynthesis back to the roots to help them grow. Because they contained networks of veins, these plants were called vascular plants. They were not very tall, as they had no good way of supporting themselves.

They still needed water in order to reproduce. But soon some of them began to parcel their offspring in tiny hard-coated spores that could be blown away by the wind. This helped the plants to spread inland to new marshy areas. Spores are often the only fossil evidence we have of these early plants.

▼ Plants first invaded land around the edges of swamps. In *Rhynia* (1), *Cooksonia* (2), and *Zosterophyllum* (3), the stems were smooth and leafless, while in *Psilophyton* (4) and *Asteroxylon* (5), they were covered with small scales. Some of the first land invertebrates were scorpion-like creatures, such as *Palaeophonus* (6), which probably evolved from the aquatic eurypterids (7). The fish were diversifying, too: Shown here are acanthodians (8); jawless armored fish such as *Pteraspis* (9) and *Cephalaspis* (10); and thelodonts (11), which were covered with scales but had no rigid internal skeleton.

THE DEVONIAN PERIOD

408 MILLION YEARS TO 360 MILLION YEARS AGO

The Devonian period was a time of great turbulence on the planet. Europe, North America, and Greenland had collided to become one large northern supercontinent, Laurasia, pushing up the ocean sediments to form great mountain ranges in eastern North America and western Europe. Erosion of the rising mountains produced huge quantities of pebbles and sand, which formed large deposits of red sandstones. As the rivers dumped their loads of sediment in the sea, vast swampy deltas grew, providing ideal conditions for animal life to take the first important steps onto land.

570 550 525 500 475 450 425 400 375 350 325 300 275 250 225 200 175 150 125 100 75 50 25 00

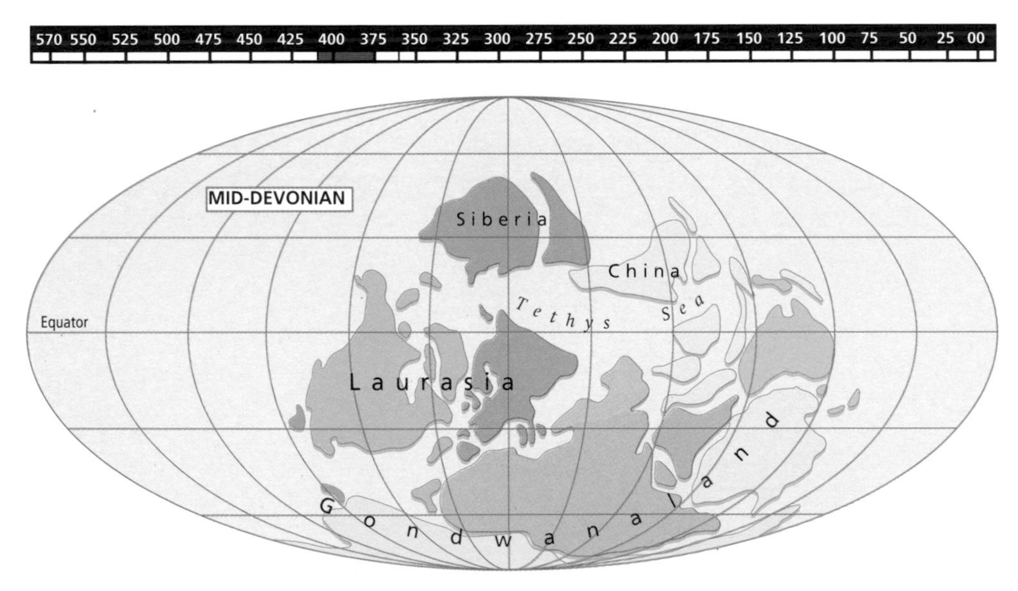

The early Devonian period saw the greatest transformation the land has ever seen. Until now there had been a landscape of bare rocks and sand – without plants to rot, there was no soil. But gradually a carpet of green plants began to spread across the land. Toward the end of the period the climate changed markedly. The earth became warmer, and severe droughts became more common, but so did periods of torrential rain. The sea level fell, and huge areas of the continents became deserts. Rivers and ponds dried up, trapping millions of fish, which provided a rich source of fossils.

The age of fish

A great variety of fish evolved during the early Devonian period. There were fish with bony armor, and fish with scales; fish with jaws, and fish with no jaws; fish with skeletons of cartilage, and fish with bony backbones. Some had fins supported by stiff rays, while others had fleshy, muscular fins.

The Devonian jawless fish (the Agnatha) had no true teeth or jaws, and their skeletons were of cartilage, not bone. Most of them, however, were covered with bony armor, and were called ostracoderms. It seems that bone arose first as defensive armor, and only later formed supportive skeletons. Many ostracoderms had large bony head shields, but as the Devonian progressed, some species arose in which the armor was simply a series of strips

interspersed with smaller scales, which allowed the fish greater flexibility for maneuvering in the water. These scales developed rather like the teeth of modern vertebrates: A soft pulp cavity was surrounded by a hard substance called dentin. Some ostracoderms had scale-covered fins and a few had a single dorsal fin (on the back) and an anal fin (behind the tail) and paired pectoral fins (behind the head), to help stabilize them while swimming.

In the bottom sediments ostracoderms with flattened bodies sucked up detritus, using their head shields to burrow in the mud. More eel-like ostracoderms swam in the open water, filter feeding or sucking in small prey. Although they had no jaws, many of these primitive fish had bony plates around their mouths, which could be moved by muscles. Most ostracoderms were small, but the heavily armored pteraspids grew up to 5 feet long.

A few jawless fish survive today. Lampreys and hagfish are long eel-like fish without a trace of bony armor, or even of bony scales. Both are flesh eaters: Lampreys

▲ A scene from the Devonian seabed. *Coccosteus* (1), a fast-moving placoderm predator, is chasing some *Tornoceras* ammonites (2), which are jet-propelling backward to escape. Ammonites and nautiloids, such as *Actinoceras* (3) and *Styliolina* (4), fed mainly on invertebrates. Trilobites like *Phacops* (5) still roamed the seabed, together with starfish (6), seen here attacking a *Camarotoechia* brachiopod (7). Many different kinds of brachiopods had evolved: *Cyrtospirifer* (8) had wings to help it keep its position in the sediment, while *Chonetes* (9), *Productella* (10), *Athyris* (11), and *Mesoplica* (12) used spines to keep themselves stable. The brachiopods and the bryozoans (sea mats) (13, 14) were filter feeders.

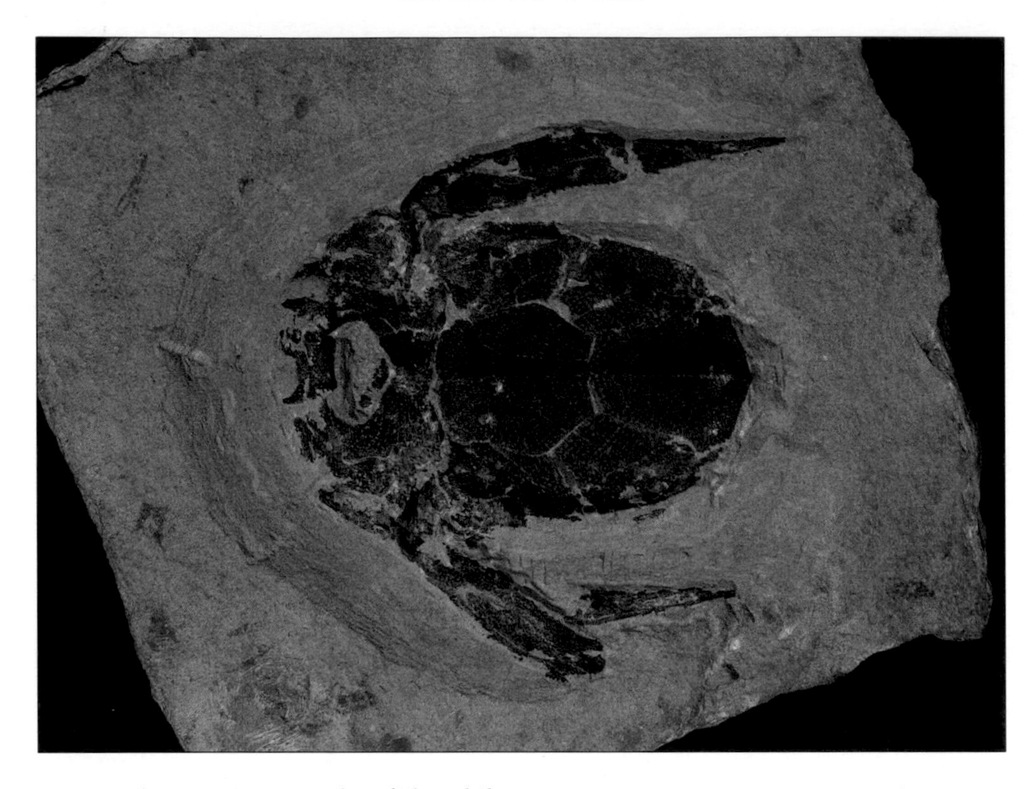

◀ Part of the fossilized head armor of *Bothriolepis*, one of a group of heavily armored fish called placoderms, the earliest fish group to have jaws. *Bothriolepis* probably scavenged on the seabed. The whole fish is illustrated on page 73.

are mostly parasites on other fish, while hagfish scavenge the corpses of other animals that sink to the ocean floor.

Jaws!

In the late Ordovician period some fish had evolved jaws and become hunters. Scientists think that some of the stiff arches supporting the gills became jaws and that the teeth evolved from the scales surrounding the mouth. One group, the placoderms ("plate-skinned fish"), included some of the largest fish in the sea, fierce predators such as *Dunkleosteus*, up to 11 feet long. Instead of teeth, they had a series of small plates in the upper jaw, which wore the edge of the lower jaw to a sharp edge that, together with the plates, could be used to bite and crush. Their massive armored heads were hinged at the neck, so the head could be thrown back as the jaws opened. The placoderms ranged far and wide through the lakes, rivers, and oceans, hunting prey no hunter before them had had the power to tackle.

But more effective predators were also evolving – the sharks. With their large fins and streamlined bodies, the first sharks glided through the Devonian seas. Their sharp teeth were always being replaced by a new set from behind. Their relatives the rays cruised over the seabed, lying in wait for unsuspecting fish and shellfish.

Old bones and new fins

However, an even more important group of fish was beginning to spread. The bony fish (Osteichthyes) – the group to which most of our modern fish belong – had arrived. Bone replaces the cartilage in their skeletons as they grow. They have two sets of paired fins – pectoral fins and pelvic fins – giving added control over twisting, turning, and braking.

The bony fish have another, very important advantage. They have a lung or swim bladder, a pouch filled with gas that helps them to adjust their density to compensate for the increase in pressure of the water at depth. By changing the amount of gas in the bladder, a bony fish can float at any depth.

From the early days of the bony fish, there were two main lines of evolution, the ray-finned fish (Actinopterygii) and the lobe-finned fish (Sarcopterygii), which are represented today only by the lungfish and the rare coelacanths. Most modern bony fish are ray-finned fish: A series of stiff rods, or rays, of bone or cartilage support their fins. The fins have no muscles of their own to move them, but are moved instead by muscles in the body wall. The lobe-finned fish have fleshy fins that are supported by a bony base. Their paired fins can be moved by means of muscles acting directly on the skeleton.

In the late Devonian, many fish groups became extinct, along with many families of corals, brachiopods, and ammonites. New species were to arise to replace them during the Carboniferous period that followed.

The greening of the land

During the Devonian period the barren land was gradually being invaded by a creeping green carpet of vegetation. The period began with a landscape of barren continents fringed with warm shallow seas and swamps, and ended with vast areas of the land surface covered in dense forest.

Some of the best evidence for early plant life comes from early Devonian fossils found around the town of Rhynie in Scotland. These plants grew in a marshy area at the edge of a small lake. They were trapped in a glassy rock called chert, which has preserved them in great detail.

Invasion of the land

Several groups of vascular (veined) plants already existed. The commonest was a plant called *Rhynia*, named after the town. It had a creeping root in the mud, from which branched several short stems no more than 7 inches tall. These stems had no leaves, and at their tips were round cases full of spores. This group of plants – called the rhyniophytes – were the ancestors of the ferns, horsetails, and flowering plants.

Another group of early plants gave rise to the lycophytes, which produced the modern club mosses. The stems of these plants were covered in thin, overlapping green scales. They became larger and more important as the Devonian went on, and eventually developed into the great trees of the coal swamps, growing up to 125 feet high. Fossil lycophyte stems often have beautiful diamond patterns formed from the scars left by the leaves. They look rather like snake skin.

Taller and taller

As the land surface alongside the lakes and waterways became more densely covered

◀ A modern club moss, with forked reproductive shoots on long stalks. Note the small leaves covering the stems – fossil club moss stems (inset) have a distinctive pattern of scars left by the bases of these leaves.

in plants, light was in short supply. Plants needed to grow taller than their neighbors in order to reach more light. For this they needed extra support. Woody tissue evolved and the first trees appeared. It was also an advantage to grow faster than your neighbor. For this even more light was needed, and it was not long before broader, flatter leaves appeared. These ancient forests would have looked very unfamiliar to our eyes. The trees stood on roots that branched above the ground, and their trunks were covered not with bark, but with reptilelike scales.

▼ An Australian lungfish. Lungfish are "living fossils" – survivors from Devonian times. They live in stagnant water that is short of oxygen, frequently coming to the surface to gulp air into their "lung." They can survive long periods of drought, buried in the mud, breathing air through a tube in the mud.

The first compost

All this vegetation produced a lot of dead wood and leaves, which would have piled up and clogged the forests if it was not removed. But by this time, there were already fungi to start breaking it down. And as the plant roots worked their way into the land surface and broke it up, bacteria moved in to work on the dead material, helping to form the first soils. And the animals were moving in, too.

The arthropods move in

With such a rich new source of food it is not surprising to find that animals were soon taking advantage of the new plants. There are lots of remains of arthropods (joint-legged invertebrates), in the Rhynie chert. Tiny mites less than 1/50 of an inch long sucked at the plant sap. And tiny spiderlike animals less than 1/10 of an inch long preyed upon them. Primitive wingless insects rather like silverfish scavenged on the dead plant material. Shrimps paddled through the shallow water, feeding on the microscopic floating life that thrived on the nutrients washed in from the rotting plant debris.

From sea tyrants to land tyrants

More powerful predators – scorpions and their ancestors – were quick to follow. The ancestors of the scorpions were probably animals like the eurypterids, which had been hunting the seas and lakes since Ordovician times. Eurypterids had large shieldlike heads and segmented bodies that often tapered to end in a long narrow spine. Paleontologists think they lived on the seabed, and many of them had both walking legs and paddlelike legs for swimming. Some had limbs armed with pincers, which they held in front of them like scorpions. Good eyesight is a great asset for predators, and the eurypterids had large compound eyes. By the Devonian period some of the eurypterids had grown very large – up to 7 feet long. They must have been some of the largest predators in the sea. Certainly they are the largest arthropods ever known.

The evolution of lungs

The vast swamps that developed toward the end of the Devonian period presented certain problems for their inhabitants. Warm water cannot hold as much oxygen as cold water, and where many aquatic animals are crowded into shallow water oxygen is soon in short supply. Most primitive bony fish gulped air at the water surface. Fine blood vessels lining their throats absorbed oxygen directly from the air. These fish eventually evolved a lung that they could fill with air, and they evolved nostrils through which to breathe in the air. In most groups of bony fish, the lung was to evolve into the swim bladder, but for many swamp dwellers its value for obtaining oxygen was immeasurable.

Lungfish survive today as "living fossils." They are lobe-finned fish found only in Africa, Australia, and South America, countries that in Devonian times were all joined together in the southern supercontinent of

Gondwanaland. These fish live in stagnant shallow water, gulping air at the surface.

The rise of the amphibians

The lobe-finned fish had one pair of fins just behind the head and the other pair in front of the tail. If you watch a newt or salamander walking, you will notice that it wiggles its body from side to side, just like a fish. This is not a coincidence. The lobe-finned fish swam in a similar way, using their fins as paddles for extra "push." Living coelacanths still swim like this. In order to support their fins, the lobe-finned fish had evolved bony structures arranged in a similar way to the limb bones of modern land vertebrates. The stage was set for the evolution of the amphibians, vertebrates that spend part of their lives in water and part on land.

It is thought that amphibians evolved from a group of carnivorous lobe-finned fish called the rhipidistians. To make the leap from living in water to living on land, the amphibians had to be able to lift their bodies off the ground in order to walk effectively. The pelvic girdle linking the legs to the backbone needed to be firmly fused to the backbone. But the skull needed to become separate from the shoulders, otherwise the force of walking or paddling would jar the skull. In the water, their backbones provided a support for the muscles involved in swimming, but the animals' bodies got all the support they needed from the water. On land, there was no such support, so changes were required to prevent the body from sagging between the legs.

The bones supporting the fleshy fins of the lobe-finned fish now had a more difficult job to do. The new limbs needed to be able to swing downward at the shoulder joint. The elbow and wrist joints became more highly developed to allow the bending, thrusting, and twisting of the limbs that were needed for walking movements. The bones of the hand became more splayed out, giving it a large surface area to spread the animal's weight over the ground.

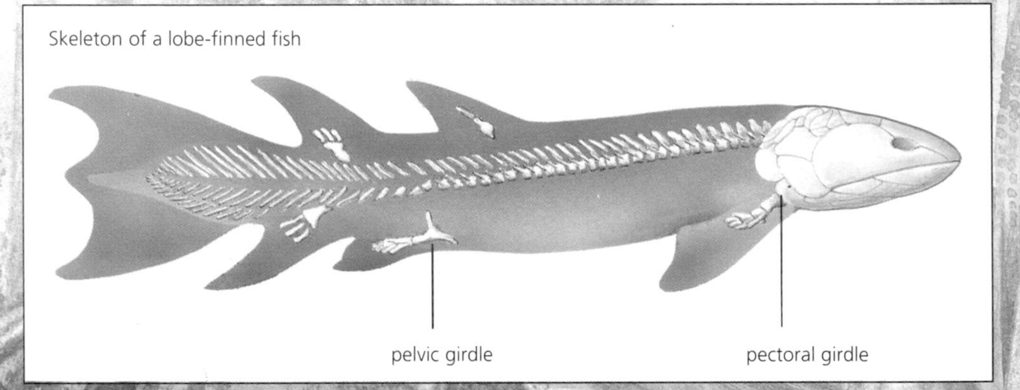

Skeleton of a lobe-finned fish

▼ Skeletons of a lobe-finned fish (left) and the first amphibian, *Ichthyostega* (right). The number and arrangement of the bones in the hind fin of the fish and the hind limb of *Ichthyostega* are very similar. In *Ichthyostega* the pectoral (shoulder) girdle joins directly with the backbone instead of being fused to the head, and the pelvic girdle has also become attached to the backbone, in readiness for supporting the body. Remains of the front foot or flipper of *Ichthyostega* have not yet been found, but its massive bones and the angle of the elbow suggest that it was rather like the front flipper of a fur seal or sea lion.

Between two worlds

The early amphibians were probably still mainly aquatic, feeding on fish and invertebrates. With their ability to breathe air, they would have flourished in swamps. However, the rapidly evolving insects offered an exciting new diet, and as yet there were no large predators on the land. Present-day amphibians still need to return to water to lay their soft eggs, and the eggs hatch into fishlike tadpoles – evidence of their fishy ancestry.

The earliest-known four-footed land animal, or tetrapod, for which we have good fossils is *Ichthyostega*. *Ichthyostega* had a streamlined body with powerful limbs, and a shoulder girdle and pelvic girdle typical of a land animal, but it also had a tail with a tail fin, and a lateral-line system (a line of sense cells used by fish to sense vibrations in the water). This shows that *Ichthyostega* still spent much of its time in the water. Its feet would rest placed flat on the ground, but with its heavy skull and ribs it would have moved only sluggishly.

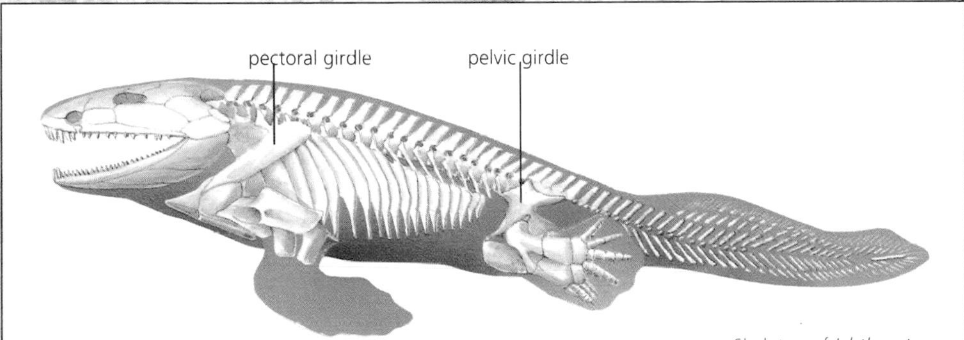

Skeleton of *Ichthyostega*

▼ A late Devonian swamp scene. Air-breathing amphibians evolved in the stagnant water of swamps. *Ichthyostega* (1) was the first known amphibian. It probably spent most of its time hunting in the water. When on land, it is thought to have used its front legs as props, rather like sea lions use their front flippers. The freshwater shark *Xenacanthus* (2) is chasing a shoal of small acanthodians (3), which are also being pursued by a bony fish, *Cheirolepis* (4). A lungfish, *Dipterus* (5), is gulping air at the surface. The placoderms *Bothriolepis* (6) and *Pterichthyodes* (7) are scavenging pieces of organic debris.

THE SEEDS OF SUCCESS

Some 3 billion years ago the first algae began to use the sun's rays to manufacture their own food, a process called photosynthesis, releasing oxygen into the earth's atmosphere. Much later in the Precambrian period, multicellular algae (seaweeds) evolved and began to clothe the seabed in shallow coastal waters. By the late Ordovician period – perhaps even earlier – these algae had spread into fresh water.

Coming out

In the Silurian period plants finally made the move onto land, evolving a waterproof outer layer, the cuticle, which was perforated by tiny pores called stomata that allowed gases in and out for photosynthesis. A network of tubes, the vascular system, developed to transport water from root to shoot, and this later became more extensive, providing woody tissues for extra support.

But it was new methods of reproduction that were the key to plants becoming established on the land. Reproduction in water is a fairly straightforward process. The male sex cells (sperms) simply swim to fertilize the eggs. This was still an option for the first land plants, which lived around the swampy edges of the water. But even early land plants, such as *Cooksonia*, were soon producing spores (reproductive cells) at the tips of their stems for the wind to distribute.

Seeds and cones

During the Devonian period plants became more complex and diverse. Ferns, club mosses, and horsetails evolved, and by the mid-Devonian plants were growing away from the water's edge. However, these ancient plants still relied on water for fertilization. It was not until the late Devonian that the first seed-bearing plants – the seed ferns – arose. The seed ferns kept their large female spores on the parent plant. The tiny male spores blew on the wind to the female spores. Only then did they release swimming male sperms. Once fertilized, a protective layer of parent tissue formed around the developing embryo, producing the first true seeds. The cycads still reproduce in this way today.

About 240 million years ago the first cones appeared. Male cones produce tiny male spores, or pollen. Female cones are usually bigger and contain the eggs. The spores are protected inside the spiral of scales on the cone. The sperms – and the

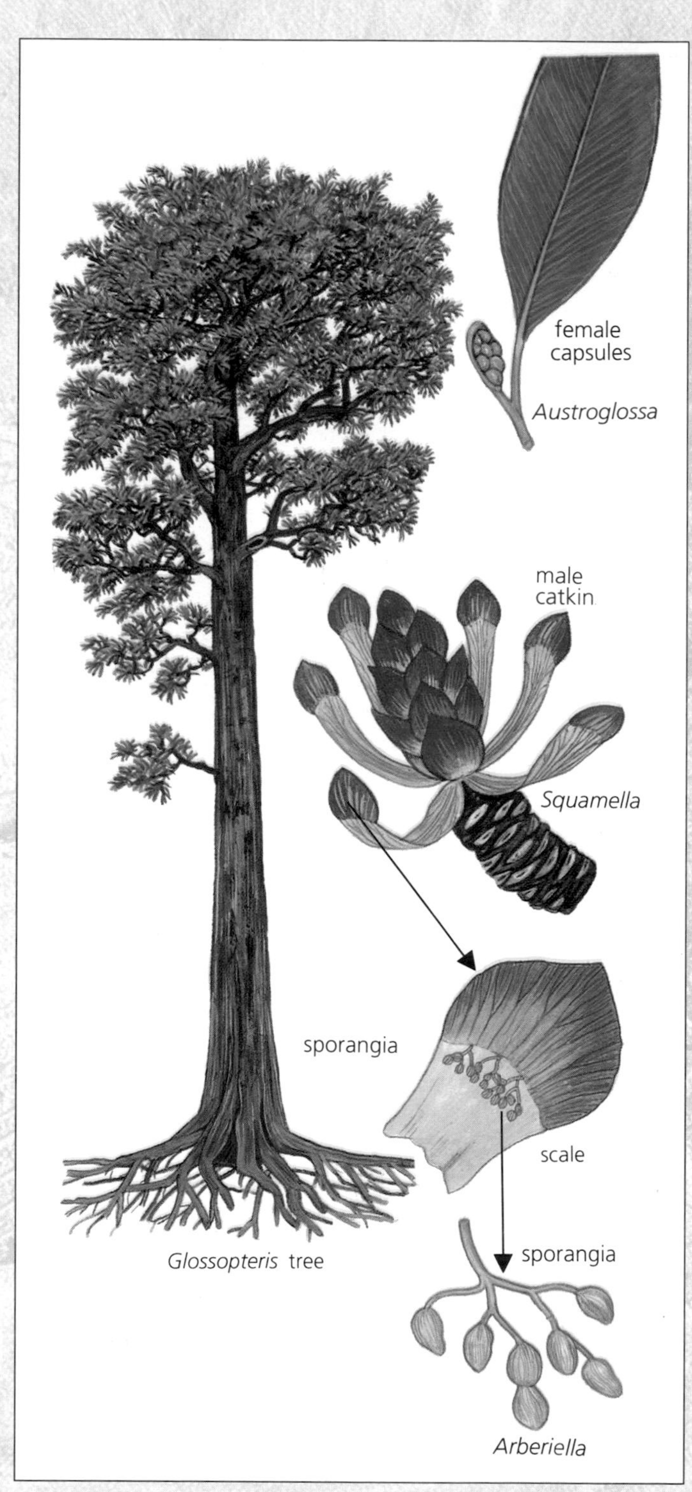

▲ A *Glossopteris* tree. The name means "tongue-frond," after its tongue-shaped leaves. Glossopterids became very common as the climate warmed at the end of the Carboniferous period. They formed huge forests across the southern supercontinent of Gondwanaland. At first the different parts were given different Latin names, until scientists realized they all belonged to the same plant. Thus *Austroglossa* was the female reproductive organs protected by a small scale leaf. When fertilized, these produced the seeds. *Squamella* was the male catkin. On the inside of each scale of a male catkin were clusters of spore capsules (*Arberiella*).

need for water – have been done away with altogether: The pollen grain grows a tube through the female spore tissue to the egg cell. Conifers are a successful design of plants – one-third of the world's forests today are coniferous.

The first flowers

During the Carboniferous period lush forests of giant club mosses, horsetails, ginkgoes, conifers, cycads, and ferns flourished. These were home to the rapidly evolving insects. The next important step forward was the appearance of the angiosperms, or flowering plants, in the late Cretaceous period. Some angiosperms evolved brightly colored petals and fragrant nectar to attract insects to transfer their pollen.

The flower has several improvements over the cone. The egg cells, and later the seeds, are produced inside the chambers of the ovary, which provides both food and protection. After fertilization, the ovary wall swells up into a fruit that gives further protection for the fertilized egg (now a seed) and the embryo plant inside it. Since the ovary can expand after fertilization, the seeds can acquire large food reserves. This enables them to germinate rapidly once they find favorable conditions.

▲ The first fern frond grows from the delicate plate of cells called the prothallium. Fern spores germinate into a moisture-loving prothallium so susceptible to drying out that it restricts most ferns to damp places. Male sex cells (swimming, spermlike antherozoids) and female ones (eggs) are produced in flask-shaped cups (antheridia and archegonia) on the underside of the prothallium. The fertilized eggs develop into the new fern frond.

New partnerships

The evolution of the fruit and its enclosed seeds coincided with the evolution of the birds and mammals. At this time the early mammals were beginning to take over from the dinosaurs. Seeds and fruits provided an enormous source of food for the up-and-coming mammals. Fruits were eaten by birds and mammals and their seeds were then distributed away from the parent plant. Some fruits had bright colors, a succulent taste, or tempting smell, to entice animals to eat them. Once swallowed, the seeds resisted digestion and passed through the gut to be deposited many miles away. The walls of other fruits developed hooks for clinging to hair or feathers, or wings to sail on the wind.

◀ The simple magnolia flower is probably very similar to the first insect-pollinated flowers. Like them, it is pollinated by beetles.

◀ The flowering plants have evolved more complicated ways of attracting insects to pollinate them. Here, a long-horned bee attempts to mate with a bee orchid flower that not only looks rather like the female bee, but also smells like her. The yellow pollen sacs of the last bee orchid he visited have stuck to his head, and their pollen will rub off on the female parts of the orchid he is now visiting.

The Carboniferous Period

360 MILLION YEARS TO 286 MILLION YEARS AGO

When the Carboniferous period began, most of the earth's land surface formed part of two great supercontinents, Laurasia in the north and Gondwanaland in the south. During the late Carboniferous, the two supercontinents moved closer and closer together. These movements pushed up new mountain ranges around the edges of the earth's plates, and great outpourings of lava covered the edges of the continents. The climate cooled, and there were at least two ice ages as Gondwanaland drifted over the South Pole.

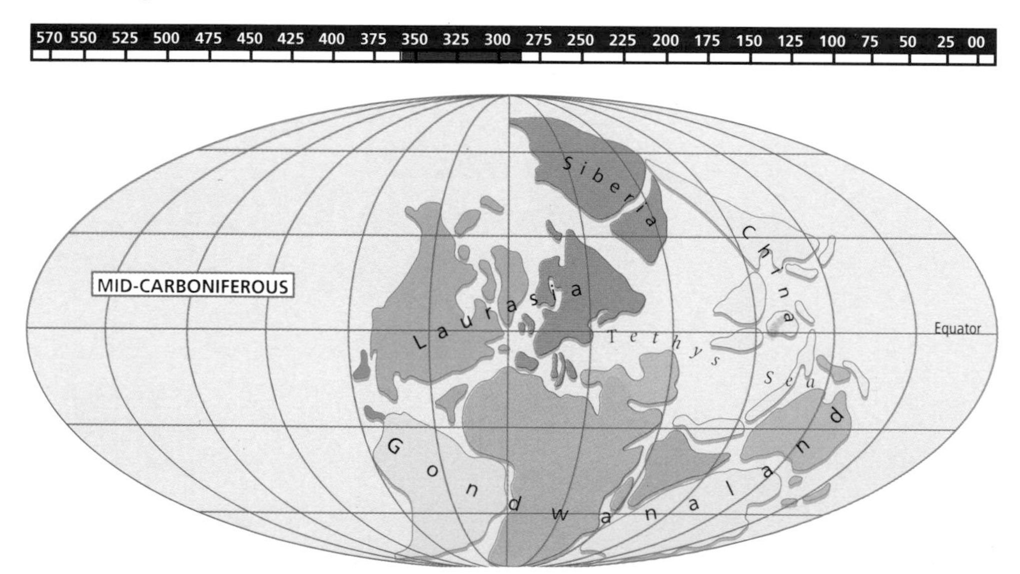

In the early Carboniferous, conditions over much of the land surface were almost tropical. There were huge areas of shallow coastal seas, and the sea invaded low-lying coastal plains to form vast swamps. In this warm, moist climate, vast forests of giant tree ferns and early seed plants flourished. They produced a lot of oxygen, and by the end of the period there was almost as much oxygen in the atmosphere as there is today.

Some of the forest trees grew up to 148 feet high. The vegetation grew so fast that there were not enough invertebrates in the soil to feed on and break down all the dead material, and it began to pile up. In the damp conditions created by the Carboniferous climate it formed thick layers of peat. In the swamps the peat was soon underwater, buried in sediment. In time these sediments formed the rocks of the coal measures – layers of sedimentary rocks containing bands of coal that had formed from the fossil plants in the peat.

Insects everywhere

The plants were not the only living things exploring the land. The arthropods had also moved out of the water, and a new arthropod group – the insects – was proving very successful. From the moment of the insects' first appearance, they have

been a success story. Today there are at least a million known species, and maybe as many as 30 million still to be discovered. Indeed, modern times could almost be called the "age of insects."

Being small, insects can live and hide in places other animals cannot reach. Insect bodies have a highly adaptable basic plan that can be adjusted for swimming, walking, running, jumping, or flying. The hard outer skeleton, the exoskeleton (made of a substance called chitin), is extended into mouthparts capable of chewing tough leaves, sucking up sap, and piercing animal skin or biting prey.

The great coal forests

The lush forests of the Carboniferous were dominated by huge tree ferns up to 148 feet high, with leaves over 3 feet long. There were also giant horsetails, club mosses, and

▲ A Carboniferous coal swamp. There are many large trees, including *Sigillaria* (1) and various giant club mosses (lycophytes) (2), and lush thickets of *Calamites* (3) and horsetails (4), an ideal moist habitat for the early amphibians, such as *Ichthyostega* (5) and *Crinodon* (6). Insects abound – cockroaches (7) and spiders (8) scurry through the undergrowth, while the air is patrolled by giant *Meganura* dragonflies (9) with a 3-foot wingspan. The rapid growth of these forests led to large accumulations of dead plant material, which sank into the swamp before it could be decomposed, and eventually turned into peat, then coal.

newly evolved seed-bearing plants. The trees had very shallow roots, often branching above the ground, and they grew very close together. There would have been a lot of fallen trees and dead wood and leaves around. In these dense jungles, growth was so fast that the decomposers (bacteria and fungi), which rot away the dead material on the forest floor, could not keep up.

These forests were warm, steamy places where the air was laden with moisture. There were many pools and swamps, breeding grounds for a huge number of insects and for the early amphibians. The air buzzed with insects – cockroaches, grasshoppers, and giant dragonflies with wings up to 27 inches long – while silverfish, termites, and beetles lived in the undergrowth. Spiders were already on the move, and millipedes and scorpions scurried across the forest floor.

Plants of the coal swamps

The plants of these great forests would look rather strange to us. The ancient lycophytes, relatives of our club mosses, formed trees 148 feet tall. Below them, up to heights of 66 feet, were giant horsetails, strange plants with rings of narrow leaves springing from thick jointed stems. There were also tree-sized ferns.

Like their descendants today, these ancient ferns could live only in damp surroundings. Ferns reproduce themselves by producing hundreds of tiny hard-coated spores that are wafted away on air currents. But before these spores can produce a new fern plant, something special has to happen. The spores germinate into tiny delicate "gametophyte" (sex-cell producing) plants that produce little cups containing male or female sex cells (sperm and eggs). The sperm need a film of water in which to swim to the eggs to fertilize them. Then a new fern plant, the "sporophyte" (spore-bearing) stage of the life cycle, can grow up from the fertilized egg.

The seed plants

The delicate gametophyte plants could survive only in very damp places. But toward the end of the Devonian period, one group of plants – the seed ferns – found a way around this. The seed ferns were similar to modern cycads or tree ferns, and reproduced in the same way. They kept the female spores on the parent plants, where they produced small flasklike structures (archegonia) containing egg cells. Instead of swimming sperm, the seed ferns produced pollen, which could be carried on air currents. These grains of pollen

HOW COAL IS FORMED

The lush Carboniferous forests grow so fast that there is not time for all their dead leaves, twigs, and tree trunks to rot. In these "coal swamps," layers of dead plant material form waterlogged peat, which turns into coal under pressure.

The sea invades, depositing the remains of sea creatures and layers of mud, which turn into shale.

The sea retreats, and rivers deposit sand to form sandstones.

The land becomes more marshy, and mud is deposited, ready to form silty sandstone.

The forest grows again, forming a new coal seam. These sequences of coal, shale, and sandstone are called the coal measures.

► *Meganeura* dragonflies were the largest that have ever lived. The steamy coal forests and swamps were home to many smaller flying insects that provided them with easy prey. Dragonflies' huge compound eyes give them almost all-around vision, enabling them to detect the slightest flicker of movement from potential prey. With this obviously successful design for aerial hunting, dragonflies have changed little in hundreds of millions of years.

◀ Part of a fossil fern, *Alethopteris*, from the coal measures. Ferns thrived in the moist, humid coal forests, but they were ill-adapted to cope with the drier conditions that followed in the Permian period. When fern spores germinate, they produce a thin, delicate plate of cells called a prothallium, which eventually bears the male and female reproductive organs. This prothallium is extremely susceptible to drying out. Furthermore, the male sex cells, or spermatozoa, produced by the prothallium have to swim through a film of moisture to reach the female egg cells. These limitations have restricted the ferns to the moist habitats they still inhabit today.

germinated on the female spores and released the male sex cells, which then fertilized the egg cells. At last the drier parts of the continents could be colonized.

The fertilized egg developed inside a cuplike structure called an ovule, which developed into a seed. This contained food stores, so that the new seedling could germinate quickly. In some plants huge cones up to 27 inches long carried the female spores and produced the seeds. This process did away with the need for water to transfer the male sex cells (gametes) to the egg cells, and also avoided the delicate, vulnerable gametophyte stage.

The age of amphibians

The early amphibians had bulging eyes and nostrils on the top of broad, flat heads. These would have been useful while swimming at the surface of the water. Some may have lurked half underwater, waiting to ambush their prey, much as crocodiles do today. These amphibians would have looked rather like giant salamanders. They were hunters, with hard, sharp teeth for attacking their prey. These teeth were easily preserved as fossils.

The amphibians soon evolved into many different forms, some up to 26 feet long. While the larger amphibians still fished in the water, the smaller ones (the microsaurs) were taking advantage of the flourishing insect life on land. Some of them had very tiny legs or even no legs, rather like snakes without scales. They were probably burrowers. The microsaurs were rather like little lizards, with short teeth for cracking open the bodies of insects.

▶ Insects and amphibians flourished in the warm swamps of the late Carboniferous. Butterflies (1), giant cockroaches (2), dragonflies (3), and mayflies (4) fluttered among the trees; giant millipedes (5) feasted on the rotting vegetation; and centipedes (5) hunted on the forest floor. *Eogyrinus* (7) was a large (15 feet long) amphibian that probably hunted like an alligator, while *Microbrachis* (8), only 6 inches long, fed on tiny plankton animals. *Branchiosaurus* (9) was a tadpolelike amphibian with gills; while *Urocordylus* (10), *Sauropleura* (11), and *Scincosaurus* (12) were more like newts; and the legless *Dolichosoma* (13) was snakelike.

The first reptiles

By the late Carboniferous, a new group of four-legged animals was wandering in the forests. Most of them were small, and looked rather like present-day lizards. This is not surprising, for these were the first reptiles. They had a more waterproof skin than the amphibians, and were able to spend all their time out of water. There was plenty to feed on: Worms, millipedes, and insects were there for the taking. And before long, larger reptiles were evolving to feed on the smaller reptiles.

A very private pond

The reptiles no longer needed to return to water to reproduce. Instead of laying soft eggs that produced swimming tadpoles, these animals laid eggs with tough leathery shells. They hatched into miniature versions of their parents. Inside each egg was a little sac full of water for the embryo reptile to live in, another sac full of yolk for it to feed on, and a third sac in which its waste collected. This liquid cushion also protected the young inside from bumps and bruises. The yolk provided food, and by the time the baby reptile hatched it no longer needed a pond (sac) to swim in. It was big enough to fend for itself in the forest.

▼ A Nile crocodile embryo inside the egg. Such eggs, which are resistant to drying out, provide protection against knocks, and contain a plentiful supply of food in the form of yolk. These features enabled the reptiles to become fully independent of the water.

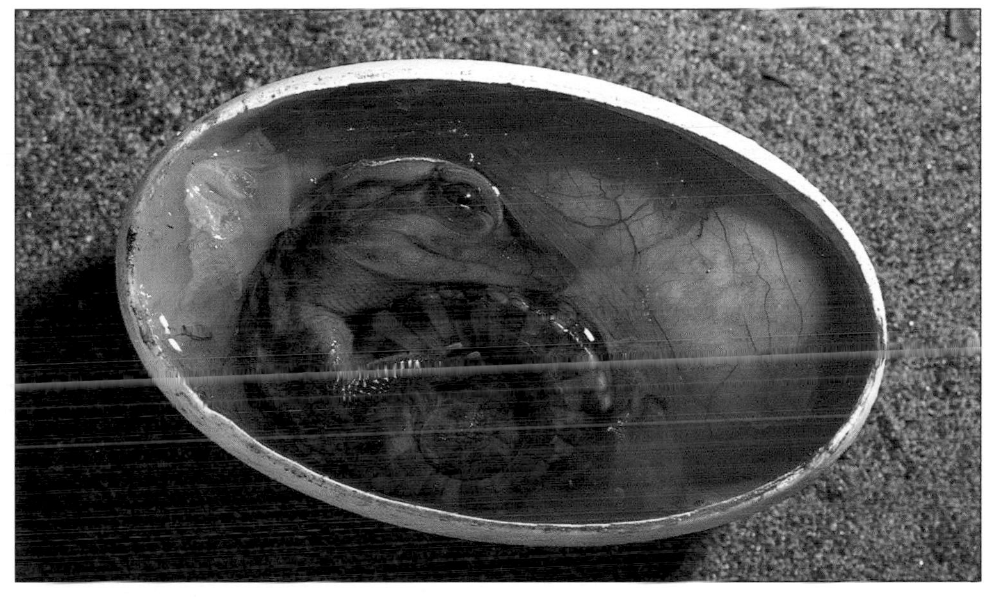

THE FIRST FLIGHT

The Carboniferous insects were the first animals to take to the air, 150 million years before the birds. The dragonflies were the first. They soon became the supreme fliers of the coal swamps. Some had wingspans of over 27 inches. They were soon followed into the air by butterflies, moths, beetles, and grasshoppers. But how did it all start?

In damp parts of the kitchen or bathroom you may have seen the little insects called silverfish (right). There is one kind of silverfish that has a pair of tiny flaps sticking out from its body. Perhaps an insect like this was the ancestor of flying insects. Maybe it spread its flaps in the sun to help it warm up in the early morning. Flapping them up and down would have warmed it up faster, rather like running on the spot warms us up. As the flaps got bigger, they may have been used to glide from tree to tree, perhaps to escape from predators such as spiders.

The Permian Period

286 MILLION YEARS TO 248 MILLION YEARS AGO

Throughout the Permian period the supercontinents of Gondwanaland and Laurasia were moving closer together. Asia collided with Europe, pushing up the Ural mountains; India reached Asia and the Himalayas were born; and the Appalachians rose up in North America. By the end of the Permian the giant supercontinent of Pangaea was complete.

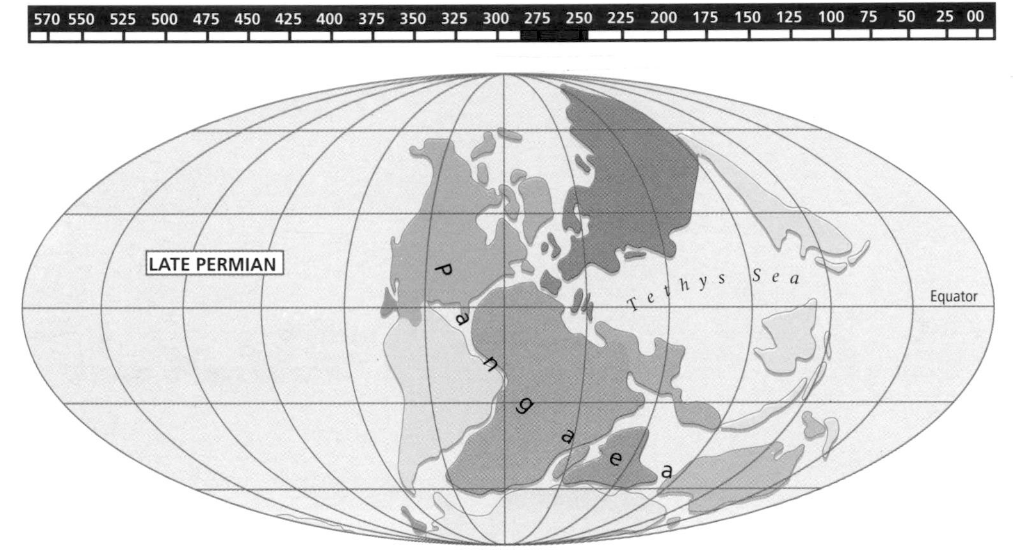

As the pattern of land and sea changed in the Permian, so did the climate. The period began with an ice age on the southern continents and a falling sea level worldwide. But as Gondwanaland moved north, the land warmed and the ice melted. Meanwhile, parts of Laurasia became very hot and dry, and deserts spread.

Life in Permian seas

During the Carboniferous the crinoids had become common on the reefs, forming strange armor-plated underwater gardens. There were lots of different kinds of brachiopods. Many had zigzag-edged shells, which made it easier to lock the two halves of the shell together. Spiny brachiopods lived in the mud, and brachiopods on stalks attached themselves to any solid object they could find, including the shells of other animals.

But they were competing for food with the newly evolved bivalve mollusks, ancestors of modern clams and mussels. Some bivalves moved into a new habitat, the sediments. They had a strong muscular "foot," which they used to bury themselves. They fed through tubes pushed up to the surface. Like modern scallops, a few species could even swim by clapping their shells together to propel themselves along.

Coiled carnivores

During the Carboniferous period, some new predators had appeared in the sea. These were the ammonites, relatives of the nautiloids. Most ammonites probably hunted just above the seabed, but a few ventured into open water. Their powerful jaws could make quick work of trilobites and other shellfish. Ammonites are very attractive fossils to collect. Their shells were decorated with ridges and knobs, and between their chambers they had plates that showed up as patterns of grooves on the outside of the fossil shell. The plates came in a great variety of patterns, and the grooves became more wiggly and frilly as the Permian period went on.

Decline of the amphibians

In the early Permian, the amphibians dominated both land and fresh water. One of the most powerful hunters, *Eryops*, was over 7 feet long. *Eryops* preyed on smaller amphibians and reptiles, and probably fish, too. Some of the strangest predators were *Diplocaulus* and *Diploceraspis*, flattened animals with huge boomerang-shaped heads. Their eyes pointed upward, so they probably lurked in the mud at the bottom of ponds, waiting for their prey to swim overhead. Nobody really knows the reason for the odd-shaped head. Maybe it was used in fights, to take sideswipes at rivals. Or perhaps it served as a hydrofoil, helping the fish to lift itself in the water while swimming.

lift

reduced pressure

water flow

▲ A possible explanation for the bizarre boomerang head of *Diplocaulus*, an early amphibian from the Midwest, is that the shape of the head may have helped to generate lift while swimming, much as the specially shaped wing of a bird or airplane generates lift for flying. When swimming against the current, the water splits as it meets *Diplocaulus*'s head. Since the top of the head is convex, the water flowing over it has to travel farther than the water below it, so it moves faster. This reduces its pressure, generating lower pressure and lifting the head up. This would have enabled the animal to swoop rapidly upward and take its prey unaware from below. To sink to the bottom, *Diplocaulus* had only to point its head downward.

▲ A glimpse of a Permian reef. The recently evolved ammonites are becoming important predators, but even more threatening are the sharks. Here, *Hybodus* (1) has seized a ray-finned bony fish, *Platysomus* (2), while its companion makes a quick getaway. The shark has frightened an ammonite (3), which is jet-propelling itself rapidly backward behind a cloud of ink it has just ejected. *Hybodus* itself is being parasitized by a lamprey, *Hardistella* (4). Filter-feeding bivalves and brachiopods, such as *Stenoscisma* (5) and *Horridonia* (6), are abundant. Much of the solid part of the reef is made up of large colonies of bryozoans (sea mats), such as *Fenestella* (7) and *Synocladia* (8), intermingled with fragments of more delicate branching bryozoans like *Acanthocladia* (9).

With such dangerous predators around, some amphibians developed their own armor. Bony plates protected their backbones, so they have been given the nickname armadillo toads. As the climate became drier, the amphibians, with their moist, porous skins, became confined to damp habitats, and many became extinct. A new group of animals, better adapted to dry conditions, began to spread across the globe: the reptiles.

The reptiles take over

The first reptiles were small lizardlike animals that fed mainly on arthropods and worms. But soon large reptiles evolved to feed on smaller ones. In time both predators and prey evolved larger and more powerful jaws for the fight, and stronger teeth firmly set in sockets (like the teeth of modern mammals and crocodiles). The reptiles were growing bigger and fiercer.

Some reptiles, including the mesosaurs, went back to the water. The mesosaurs had needlelike teeth that interlocked when the jaws closed. They acted like a strainer. The mesosaur took in a mouthful of small invertebrates or fish, strained out the water through its teeth, and swallowed the solid remains.

By the end of the Permian period, a group of faster moving, mammallike reptiles – the gorgonopsians – had evolved. The early reptiles still had their legs at the sides of their bodies, like many lizards today. They could only waddle, and their bodies twisted from side to side as they went. But the new gorgonopsian reptiles had their legs farther under their bodies. This enabled them to take longer strides and run faster. Many gorgonopsians had huge fangs that could pierce the tough skins of armored reptiles.

Plant-eating reptiles

The mammallike reptiles, or synapsids, had evolved during the late Carboniferous period. The most primitive of them, the pelycosaurs, had evolved into many different species to become the largest and commonest reptiles. Most pelycosaurs had large teeth, which suggests that they fed on large prey. Some species became adapted for eating plants. Plants take a long time to digest, so these animals needed to keep lots of food in their stomachs for long periods of time. To do this they needed to be bigger. But it was not long before the carnivorous reptiles (the predators) grew bigger, too.

In the late Permian, other groups of mammal-like reptiles arose, such as the dicynodonts. Some were the size of a rat, while others were as big as a cow. Most lived on land, but a few became aquatic. The dicynodonts had teeth in sockets, though most only had a pair of large canine teeth for biting plants. Dicynodonts probably had horny beaks like those of tortoises. Some had tusklike teeth. They probably used them to scrape for roots.

The great extinction

By the end of the Permian, the northern land masses were very arid. On the fringes of the swamps and lakes conifers were thriving, along with tree ferns, ferns, club mosses, and a few horsetails. The southern continent was not so dry, and was still separated by ocean from the north. Many of the old plants had been wiped out by the earlier ice ages, and forests of *Glossopteris* had taken over. *Glossopteris* produced seeds, and may well be the ancestor of the present-day flowering plants.

The late Permian was a time of great upheavals. Continents were colliding, mountains were rising, the sea was advancing and retreating, and the climate was changing. Millions of animals and plants failed to adapt to all these changes and became extinct. In the worst extinction in the history of the world, more than half of all animal families disappeared. The species living in shallow water were worst hit – more than 90 percent of them perished, including more than half the amphibians and most of the ammonites. The old wrinkled corals passed away, and were replaced by modern reef-building corals. And the trilobites finally became extinct.

THE SAIL-BACKED REPTILES

The sail-backed reptiles were a bizarre group of pelycosaurs. Some, such as *Dimetrodon*, grew very large (over 11 feet long). Running along their backs were huge sails supported by long spikes growing out from the backbone. These were probably used to control their body temperature. The sails were well supplied with blood vessels. In the cool of the early morning, the sail backs turned their sails to face the sun, so they could warm up and become active quickly. Once they had warmed up, they could easily attack other reptiles that were still cold and sluggish. When it became too hot, they turned around, so that only the thin edge of the sail faced the sun.

The seeds of destruction

Many reasons have been put forward for the scale of the Permian extinction. Many species lost their habitats as mountain barriers arose and seas, lakes, and rivers disappeared. Some were unable to cope with the change in climate as the continents migrated northward. As the continents merged, their species were free to mingle, so there was a lot more competition and some species were bound to lose out.

In particular, large numbers of species that lived in fresh water and in the oceans disappeared. We can only guess why. As the climate became drier, more water evaporated from rivers and lakes, so they became saltier. Salt deposits are found in Permian rocks today. Perhaps the amount of salt in the seas, lakes, and rivers varied a lot, and many creatures could not cope.

Another great leap forward

Toward the end of the Permian period, some groups of reptiles became warm-blooded. This meant that they could stay active longer, and did not have to wait to warm up at the start of the day. To keep up their body temperature, they needed to process their food more quickly to release the heat energy from it.

One group of warm-blooded synapsids, the cynodonts, evolved a variety of different kinds of teeth, just as mammals have today. They had sharp chisel-like front teeth (incisors) for grasping and biting food, fanglike canines for stabbing and ripping at flesh, and flat molars with many cutting edges for grinding and chewing. Their skulls were also adapted to allow powerful muscles to be attached for chewing. As in crocodiles, a platelike structure called a palate separated the cynodonts' nostrils from their mouths. This enabled them to breathe through their noses while they still had food in their mouths, so they could chew their food more thoroughly. Whiskers probably grew in tiny pits on each side of the cynodonts' snouts. Scientists think the cynodonts had fur to help keep them warm. They were remarkably like mammals.

Mammals in waiting

But just as the cynodonts were becoming successful, a new and more fearsome group of reptiles appeared on the scene: the dinosaurs. Under attack from the dinosaurs, only some small, warm-blooded species of cynodonts survived. This was because they could stay active even in the cold, which meant that they were able to feed at night, when the great dinosaurs were inactive. Most of the cynodonts became extinct at the end of the Permian period, but a few survived into the Triassic. Their descendants were to survive the age of dinosaurs and give rise to the new masters of the earth, the mammals.

▼ Reptiles dominated the arid Permian landscape of southern Africa. There are mammal-like reptile predators – *Lycaenops* (1) attacks the slow-moving amphibian *Peltobatrachus* (2), despite the latter's body armor, while *Titanosuchus* (3) stalks the mammallike grazing reptiles *Moschops* (4) and *Aulacocephalus* (5). Lizardlike reptiles include *Coelurosauravus* (6), whose winglike rib flaps have a total span of 12 inches, and *Thadeosaurus* (7). *Claudiosaurus* (8) is an amphibious reptile, while *Mesosaurus* (9) is truly aquatic.

The Triassic Period

248 MILLION TO 213 MILLION YEARS AGO

The Triassic period was the beginning of the Mesozoic, or "middle-life era," in the history of Earth. All the continents had been joined together in the great supercontinent Pangaea, but this was now beginning to break up. The climate was similar worldwide. Even the temperature at the poles and the equator was much closer than it is today. Later in the Triassic, it became drier. When this happened, huge deserts developed inland as lakes and rivers began to dry up.

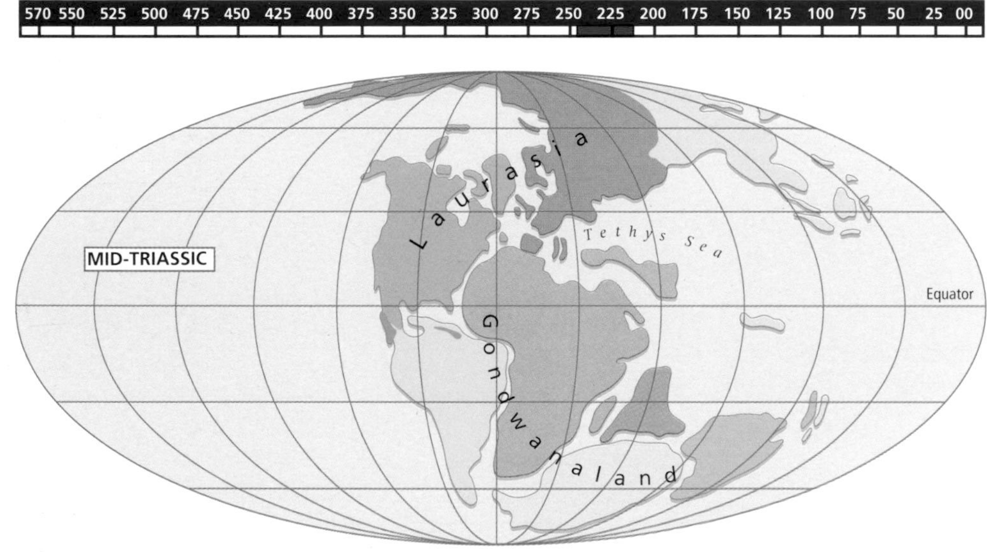

When the continents joined up to form Pangaea in the Permian, many of the world's coastlines disappeared as the great landmasses crunched up against one another. The warm climate in the Triassic then caused some of the shallow seas that were left to evaporate, and this made the remaining water very salty. Many of the old species of marine life became extinct and new types replaced them.

Invertebrates galore

New shellfish such as oysters evolved. Mollusks buried in the sand filtered food particles from the water passing over them. There were also plenty of new gastropod mollusks (snails and their relatives). As the sea level fell in the shallow seas, rocky shores were exposed. These became home for new species of mollusks such as limpets, periwinkles, and top shells. There were also new types of coral, shrimps, and lobsters.

► In the Triassic period, the continental regions were hot, dry places inland. Here, vast barren deserts formed where very few plants could grow. However, pockets of fertile land rich in plant life could be found nearer the coasts.

7
5
8
9
6
6
2
Key
1 Lystrosaur
2 Rhyncosaur
3 Gingko
4 Monkey-puzzle
5 Yew
6 Cycad
7 Tree fern
8 Bennettitalean
9 Club moss
10 Horsetail
11 Fern

The Triassic also saw the appearance of the first "modern" sea urchins. Ammonites were still around. They almost died out at the end of the Triassic, but a few survived into the Jurassic when they flourished again.

From sharks to fishing rods

Farther out at sea there were the latest kinds of fish. Sharks and bony fish hunted for food. They had evolved jaws that enabled them to crunch crabs and shellfish, such as mussels.

The biggest predators in Triassic seas were newly evolved reptiles. Lizardlike nothosaurs caught fish with their sharp teeth. Dolphinlike ichthyosaurs used sheer speed to overtake their prey. Large newtlike placodonts grubbed around on the sea bottom, picking off shellfish and then crushing them with their powerful flat teeth.

Tanystrophaeus had a long, thin neck that was twice as long as its body. It was a land animal, so it probably used its slender neck like a fishing rod. It could stand on the water's edge and reach fish swimming well below the surface offshore.

Triassic herds

At the start of the Triassic, land animals were similar the world over. Species could easily travel across the whole of Pangaea because there were no large oceans to stop them from wandering. Many animals that lived in the Permian became extinct at the start of the Triassic, probably because of climatic changes. But some of the mammal-like reptiles survived, many of them in large numbers. Great herds of plant-eating *Lystrosaurus* wallowed on the edges of lakes and rivers. They were the "hippopotamuses" of the Triassic world. Their fossils have been found as far apart as China, India, South Africa, and even Antarctica. Early in the Triassic, the first frogs hunted alongside them. Later, these were joined by the first tortoises, turtles, and crocodiles. It was not long before turtles and crocodiles invaded the sea. There, they quickly spread themselves around the world in the warm oceans.

The "dog-toothed" and ruling reptiles

There were still some cynodont ("dog-toothed") reptiles, fast-running predators, around to prey on the slow-moving herds of plant eaters. There was also a new group of reptiles called archosaurs, the "ruling lizards." When they first evolved, they were small animals that spent their time hunting around the edges of lakes and rivers. Later they evolved into much bigger animals.

In the middle of the Triassic a new group of reptiles appeared that was related to the archosaurs. They were the plant-eating rhynchosaurs, the "beaked lizards." These herbivores had a strange beak on the end of the snout that they used like a pair of tongs to gather food. Their jaws and teeth were also designed for cutting and chopping.

► *Cynognathus* was a wolf-sized cynodont ("dog-toothed") reptile. It was a powerful animal with many mammal-like features.

Scientists are sure it had hair because they have found whisker pits on its fossilized snout. This suggests these animals were probably warm-blooded, as the possession of hair is always associated with warm-blooded mammals.

When the mouth was shut, the lower jaw fitted into a groove in the upper jaw just like the blade of a penknife fits into the handle when it is closed.

From thecodont to dinosaur

Toward the end of the Triassic, many of the land animals that evolved at the start of the period died out. Now new reptiles evolved to take their place. About 225 million years ago, a group of reptiles called thecodont ("socket-toothed") reptiles appeared. At first these were clumsy, sprawling animals that looked a little like crocodiles. They lived in water and swam by waggling their powerful tails and kicking with their back legs, which were much bigger than their front limbs. When early thecodonts left the water and came onto land, their strong back legs soon became adapted to moving on solid ground.

Thecodonts quickly became efficient walkers and runners. For most of the time on land, they were four-footed. But they could also turn themselves into sprinters. It was easy for them to do this. All they had to do was a kind of thecodont "wheelie." They leaned back on their extra-big hind limbs and became two-legged runners, using their long tail as a counterbalance. Within another 20 million years, the thecodonts had developed into the first dinosaurs.

◀ The first dinosaurs were small, slender animals. At first many of them had a similar shape – they looked more like birds than dinosaurs. *Saltopus* ("leaping foot") was no bigger than a cat, *Halticosaurus* measured nearly 20 feet from head to tail, and there were all kinds of sizes in between.

Two more important "firsts"

Toward the end of the Triassic, there were two more important evolutionary developments. One of these took place on land when the first mammals appeared. The other development happened in the air with the arrival of the pterosaurs ("winged lizards").

Pioneers in flight

Some animals had already tried to fly. A small lizard called *Weigeltisaurus* of the Permian period was one of the early experimenters. But it did not have real wings. Instead, it glided from tree to tree on wings stretched between enormously long ribs. Pterosaurs improved on this design and became the first vertebrates to evolve into real heavier-than-air flyers. The new pterosaurs evolved a different wing structure that allowed them to become much better flying animals.

▼ *Icarosaurus* was a lizardlike animal that lived in North America in the early Triassic. It was a "swing-winged" glider. Its wings were formed of skin stretched tightly over long ribs. It climbed around in trees with its wings folded against its body. When it launched itself into space, its ribs swung forward and its wings opened. They worked like parachutes to slow its descent as it glided back to earth.

The Amazing Ammonites

The ammonites, which first appeared in the Devonian period, are among the best-known and most numerous of marine fossils. They belonged to the cephalopod group of mollusks and were ancestors of today's octopuses and squids. They reached a peak in terms of numbers and species in the Permian. Then, 245 million years ago, they almost disappeared in the mass extinction that took place at the end of this period. But the ammonites were not quite finished. Some survived into the Triassic, where, helped by their ocean-going capabilities, they soon spread around the world once more, reaching a second climax of evolutionary success in the middle of the Mesozoic. In fact, ammonites were so common in Mesozoic seas, and their fossils are so abundant in rocks of the time, that they have become important in determining a system for identifying marine sediments of the Mesozoic era. However, their success could not last and, at the end of the Cretaceous period, they suddenly became extinct along with many other marine animals, including the belemnites, pliosaurs, ichthyosaurs, and plesiosaurs.

From straight shells to coils

The first cephalopods, called nautiloids, developed long, conical shells, with internal partitions separating a series of gas chambers. These animals had a primitive way of drawing in and expelling water, and they used the outflow to give them a kind of jet propulsion. Over millions of years since then, cephalopods – ammonites included – perfected this technique as their main means of swimming. During the Paleozoic era (570-225 million years ago) nautiloids were the main predators of the oceans. Later, more advanced cephalopods evolved, including the ammonites, which developed a new kind of coiled, flat shell that was often highly sculptured.

▼ This diagram shows the inside of a nautilus. Ammonites probably had a similar internal structure.

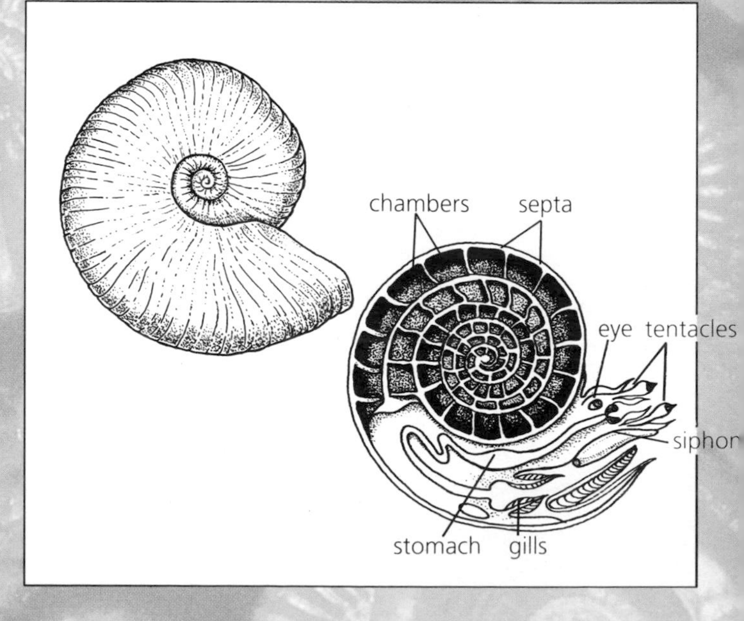

Septa and sutures

Like the modern nautilus, an ammonite's shell was divided into a series of internal chambers, each one separated from the next by a division called a septum. As in the case of the nautilus, the living animal lived in the last and most recently formed chamber. The position where each septum attached to the shell is often visible on the fossil remains of ammonites. This is called the suture line. These lines came to form a very

This typical ammonite fossil shows the coiled shell in a distinctive flat spiral shape. The surface of the shell is marked with ornamental "ribs" that indicate the positions of the septa. When you look at an ammonite fossil, it is hard to imagine that it is the remains of an animal that is closely related to today's octopuses and squids. But comparison with another living cephalopod mollusk, the nautilus, quickly makes the fact believable.

complex pattern on ammonite shells from the Jurassic and Cretaceous periods, and they have been used by scientists as a means of classifying the enormous number of ammonite fossils discovered over the years.

Ballast and buoyancy

The ammonite shell functioned as a ballast (stabilizing) organ in the same way that the nautilus uses its chambered shell today. The animal was able to fill and empty the chambers behind it with water through a structure called a siphon. The taking on board and pumping out of water allowed the animal to vary its buoyancy rather like a submarine. When an ammonite wanted to dive deeper, it filled its "ballast tanks." When it wanted to surface or float higher in the water, it emptied them.

Disappearance of the ammonites

The great abundance of ammonite fossils has been a little misleading and has led earlier scientists to make assumptions about their distribution that may not have been true. Until quite recently, scientists thought that ammonites lived in all the prehistoric seas. But then it was realized that not all species were spread around the world's oceans. Different species had different distribution patterns, and these were probably linked to physical factors such as water temperature and salinity (salt concentration). When continental movements took place at the end of the Cretaceous period, there must have been great environmental and climatic changes that affected the earth's oceans. The ammonites were probably unable to adapt to these new conditions and therefore became extinct.

Ammonites developed shells with various degrees of coiling. Usually this was in the form of a single coil, but later some species evolved partly uncoiled shells shaped a little like a wiggly question mark. Still later models carried shells shaped more like that of a snail.

The Jurassic Period

213 MILLION TO 144 MILLION YEARS AGO

By the start of the Jurassic, the gigantic supercontinent of Pangaea was well on the way to breaking up. There was still a single, large continent south of the equator once again called Gondwanaland. Later, this would also split to form present-day Australia, India, Africa, and South America. Land animals were no longer able to travel quite so easily in the Northern Hemisphere, but they could still move freely across the southern continent.

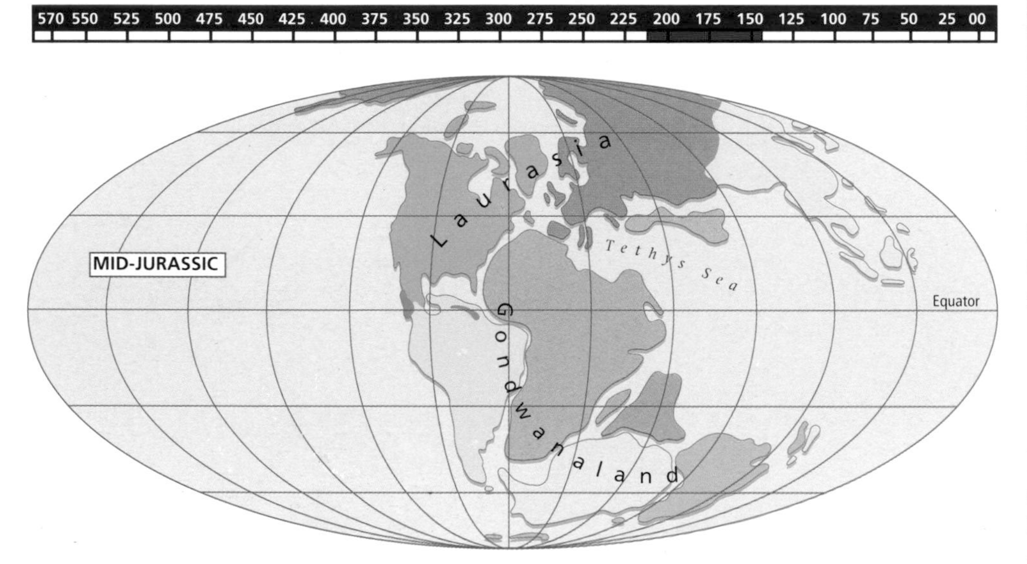

The world's climate was warm and dry at the beginning of the Jurassic. Then as rains began to soak the old Triassic deserts, there was a return to a greener world with more luxuriant vegetation. The landscape became covered with plants such as horsetails and club mosses that survived from the Triassic. The palmlike bennettitaleans also survived, and there was plenty of fungi. Forests of seed ferns, ferns, tree ferns, and fernlike cycads now spread away from the wet riverbanks to cover more of the land. Conifer forests continued to grow, with ginkgoes and monkey-puzzles. They also included the ancestors of today's cypresses, pines, and redwoods.

Life at sea

As Pangaea began to break apart, more seaways opened up in which new types of animals and algae found a home. Additional sediments gradually built up on the seabed. These were invaded by invertebrates, such as sponges and bryozoans (sea mats). Other things were also happening in the warm, shallow seas. Great coral reefs sprang up, which were home to more ammonites and new types of belemnites (early relatives of today's octopuses and squids).

Many species of crocodiles now lived on land and in lakes and rivers worldwide. There were also sea-going crocodiles with long snouts lined with sharp teeth for catching fish. A few types even swam using paddles instead of feet. They had tail fins to help them swim faster. There were also new kinds of turtles. More species of plesiosaurs and ichthyosaurs evolved to compete with the latest fast-swimming sharks, and new types of highly mobile bony fish.

▲ This cycad is a living fossil. It is almost identical to its relatives that grew in Jurassic times. Cycads are now found only in the tropics. However, 200 million years ago they were much more widely spread.

Belemnites – the bullet animals

Belemnites were closely related to today's cuttlefish and squids. They had an internal, bullet-shaped skeleton. The main part of this was made of chalky material and was called the guard. At the front end of the guard there was a cavity into which fitted a delicate, chambered shell designed to help the animal float in the water. The complete skeleton was housed inside the animal's soft body, where it provided a solid attachment for the body muscles.

The hard guard is the part of a belemnite's body that is usually found as a fossil. But sometimes "guardless" fossils are also found. Such finds puzzled scientists when they were first discovered in the early part of the 19th century. They thought they were belemnite remains, but they did not look quite right with no guard attached. The answer to this puzzle was quite simple once more had been found out about the feeding habits of ichthyosaurs, the main predators of belemnites. The "guardless" fossils were probably the result of an ichthyosaur spitting out the softer parts but keeping back the hard skeletal structures after gorging itself on a school of belemnites.

Like today's octopuses and squids, belemnites also produced an inky substance that they used to make "smoke screens" when trying to escape from predators. Fossilized belemnite ink sacs (structures for storing their ink) have also been found. A Victorian scientist, William Buckland, even managed to extract the ink from some fossilized ink sacs and used it to illustrate his book, *The Bridgewater Treatise*.

◀ Plesiosaurs were barrel-shaped marine reptiles with four large flippers to power themselves through the water.

This famous photograph, taken in Scotland in 1934, was recently denounced as a forgery. However, for 50 years it fueled speculation that the Loch Ness monster was a living plesiosaur.

A sticky forgery

A complete belemnite fossil (soft part and guard) has never been discovered, although an ingenious attempt to fool the scientific world with a forgery took place in Germany in the 1970s. Complete fossils from a quarry in the southern part of the country were bought by a number of museums for very high prices before it was realized that, in each case, the calcite guard had been carefully glued onto the fossilized soft parts!

Built for speed

Ichthyosaurs first appeared in the Triassic. They were reptiles that were perfectly designed for life in the shallow seas of the Jurassic. They had a streamlined body, fins, flippers, and long, narrow jaws. The largest were about 26 feet long but many species were no bigger than a human. They were fast swimmers, feeding mainly on fish, squids, and nautiloids. Although ichthyosaurs were reptiles, fossil evidence suggests that they gave birth to live young as mammals do. Baby ichthyosaurs were probably born at sea, as whales are.

Plesiosaurs were the other main predatory reptiles in Jurassic seas. Long-necked types lived near the surface. Here their flexible necks helped them catch schools of small fish. Shorter-necked species, called pliosaurs, dived below the surface. They fed on ammonites, and other mollusks. Some of the bigger pliosaurs probably also chased smaller plesiosaurs and ichthyosaurs.

Life in the Jurassic airspace

Insect evolution sped up in the Jurassic and the landscape was quickly changed into a buzzing, creepy-crawly world, as many new insect species evolved. Some of these

Mary Anning (1799-1847) was only 11 years old when she found the first ichthyosaur fossil skeleton at Lyme Regis in Dorset, England. Later, she also discovered the first plesiosaur and pterosaur fossil skeletons.

Miss Anning as a child ne'er passed
A pin upon the ground
But picked it up and so at last
An ichthyosaurus found.

▶ The information gained from analyzing fossilized ichthyosaur stomachs and droppings (coprolites) shows that fish and cephalopods (ammonites, nautiloids, and squids) formed their main diet. Stomach contents give even more interesting information. The small, hard hooks found on the tentacles of squids and other cephalopods were obviously a problem for ichthyosaurs because they were indigestible and not able to pass easily through their digestive system. Instead, they collected in the stomach where they acted as a digestive record of what an individual animal had eaten in its lifetime. One fossilized ichthyosaur showed that it had gulped down at least 1,500 squids!

▼ Ichthyosaurs were dolphin "look-alikes" apart from the tail, and an extra pair of fins. For a long time, scientists thought that all the ichthyosaur fossils discovered had broken tails. Then they realized that the backbone was actually bent in shape, and this shape supported a vertical tail fin (unlike the horizontal one found in dolphins and whales).

were the ancestors of present-day ants, bees, caddis flies, earwigs, flies, and wasps. Later, in the Cretaceous, an evolutionary "explosion" took place as insects began to associate with the newly evolved flowering plants.

Up until this time, only the insects had evolved into real flying animals, although a number of gliders had already tried to get airborne. Now the pterosaurs took to the air in large numbers. These were the first and largest flying backboned animals. Although they first appeared toward the end of the Triassic, it was not until the Jurassic that the pterosaurs really "took off." Their lightweight skeleton was made of hollow bones. The first pterosaurs had tails and teeth, but in more advanced types these structures disappeared as part of a weight-reduction exercise. Some fossil pterosaurs show signs of a hairy body. This suggests they may have been warm-blooded.

Scientists disagree about certain aspects of pterosaur biology. For example, pterosaurs were originally thought to be gliders, soaring like vultures on hot air rising off the land. They may have even glided like albatrosses on winds blowing above the surface of the sea. But now some experts suggest they could flap their wings in powered flight like birds. Some may have even walked like birds, while others shuffled along, or slept upside down in pterosaur "rookeries" like bats.

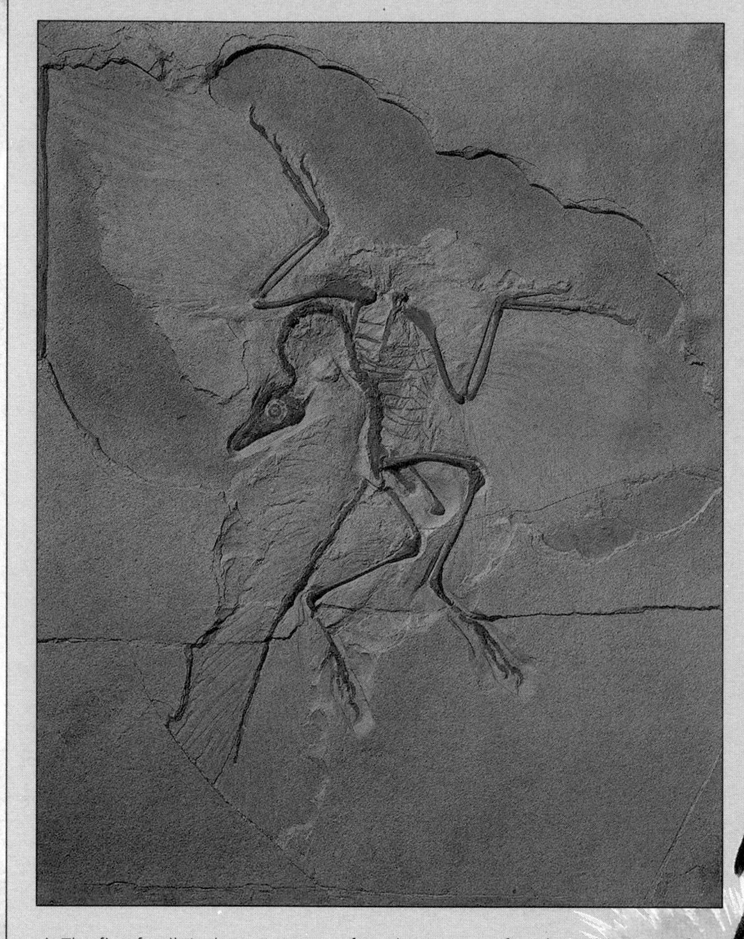

▲ The first fossil *Archaeopteryx* was found two years after the publication of Charles Darwin's book, *On The Origin of Species*. This important discovery supported Darwin's theory that evolution was a slow process and that one group of animals developed into another by means of intermediate types. The famous scientist and friend of Darwin, Thomas Huxley, predicted that an animal like *Archaeopteryx* must have existed even before its remains came to light. He actually described it in detail before its remains were discovered!

THE "EARLY" BIRD

Birds made their first appearance toward the end of the Jurassic. The first, called *Archaeopteryx*, looked more like a small, feathered dinosaur than a bird. It had teeth, and a long bony tail supporting two rows of feathers. It also had three clawed fingers jutting out from each wing. Some scientists think *Archaeopteryx* used its clawed wings to climb trees before fluttering down to earth again. Others think it took off by rushing headlong into the wind. As birds gradually evolved, their skeletons became more lightweight, and their teeth were replaced by a toothless beak. A large breastbone also evolved to anchor the powerful flight muscles. All these adaptations helped improve the basic bird body plan, and gave it the right design for efficient flight.

The first fossil record of *Archaeopteryx* was a single feather discovered in 1861. Soon after, a complete skeleton (feathers and all!) was found in the same area. Altogether, six fossil skeletons have been discovered, some complete, others only part skeletons. The most recent find was in 1988.

The evolution of flight in birds

There are two main theories about how flight evolved in birds. One suggests that it developed from the ground up. It argues that flight began as the result of a two-legged birdlike animal running and jumping up in the air. Perhaps it jumped to try and escape from predators, or perhaps it leaped to catch insects. Then, as the feathered area of the "wings" gradually became bigger, so the jumps became longer and the bird remained off the ground for longer periods. Add flapping movements, and it is easy to see how, over long periods of time, these pioneers in flight started to stay airborne as their wings gradually evolved into structures designed to support their bodies in air.

There is an opposite theory that suggests that flight developed from the trees down. The would-be flyers needed to climb to a good height before launching into the air. Gliding would have been a first step because it is a type of movement that needs very little energy, certainly less than the "running, jumping theory." Transport costs would be small because a gliding animal is pulled downward toward the earth's surface by gravity.

Stage flight

One scientist has developed an ingenious theory that imagines a series of steps which early pioneers of flight might have gone through on the evolutionary path to becoming flying animals. This theory suggests that a group of small reptiles called "protobirds" started to live in trees. Perhaps they went there because it was safer; easier to find food; or to hide, sleep, or make their nests. In the treetops, it would have been cooler than on the ground. Because of this, the reptiles evolved warm-bloodedness and featherlike structures for better insulation. Any extra, long feathers on the arms would have been useful. They would have certainly provided extra insulation but they would have also increased the surface area of the winglike arms.

But soft, feathery arms may have had another function of helping to break falls to the ground if an animal lost its balance in the treetops and plunged earthward. They would have slowed the rate of fall (the parachute effect) and also helped achieve

► Each pterosaur wing was made of skin stretched tightly between the arm bones, an enormously long fourth finger, and the body. The wings of modern-day bats are supported by four elongated fingers.

a soft landing by providing a built-in cushion. Gradually, feathered arms would have worked like prototype wings. Going through a parachute stage to a gliding stage would have been a natural evolutionary development and, finally, a flapping stage would have been developed, a step almost certainly reached by *Archaeopteryx*.

Life on land

Mammals had begun to appear in the Triassic period. The first of these were small, insect-eating animals that evolved from mammal-like reptiles called cynodonts. In the Jurassic, many new groups of mammals evolved. These included types that looked like the rats and shrews we see today. In terms of evolution, mammals played a "waiting game." Their success story was millions of years away in the future. The dominant animals in the Jurassic were the dinosaurs. Although they first appeared toward the end of the Triassic, the dinosaurs evolved rapidly in the Jurassic. They quickly established themselves as rulers of the Mesozoic world.

► In the middle Jurassic, big, heavily armored dinosaurs like *Stegosaurus* appeared. Its back legs were as tall as a room, and it had armor plates running down its back. The biggest of these plates was nearly a yard long. The best-preserved dinosaur skeleton ever discovered was a *Stegosaurus* dug up in Colorado in 1992. Scientists protected the 140 million-year-old skeleton in plaster before pulling it out of the earth by helicopter.

▼ Packs of allosaurs hunted on most continents in the Jurassic. They must have been a terrifying sight: Each member of the pack weighed more than a ton. Between them, they would easily have been able to overpower a large sauropod.

The age of dinosaurs

The earliest dinosaurs appeared more than 200 million years ago. Over a period of 140 million years, a great variety of species evolved. They spread to every continent, and adapted to a great range of habitats, although none became a burrower, a climber, a flyer, or a water dweller. Some dinosaurs were no bigger than a squirrel. Others weighed more than 15 fully grown elephants. Some lumbered along on all fours. Others ran on two legs, faster than an Olympic sprint champion.

Sixty-five million years ago, the dinosaurs suddenly disappeared. But before they vanished, they left behind a rich fossil record of their life and times.

The most common dinosaurs in the Jurassic were the prosauropods. Some of these evolved into the biggest land animals of all time, the sauropods ("lizard feet"). These were the "giraffes" of the dinosaur world. They must have spent all their time browsing on leaves from the treetops. It would have taken an enormous amount of food to fuel such a huge body. Their stomachs were massive digestive tanks for processing a continual supply of vegetation.

Later, groups of small, fast-running dinosaurs developed – the hadrosaurs. These were the "gazelles" of the dinosaur world. They cropped vegetation with their horny beaks, and then chewed it with their ridged cheek teeth.

The megalosaurids, or "great lizards," made up the largest family of big, carnivorous dinosaurs. *Megalosaurus* was a 1-ton monster with huge saw-edged teeth for cutting through meat. Some pathways show it was pigeon-toed. Perhaps it waddled along, swinging its tail from side to side like a gigantic duck. Megalosaurids wandered to all parts of the globe, and their fossil remains have been discovered in places as far apart as North America, Spain, and Madagascar.

The earliest species were probably relatively small and lightly built animals, but later megalosaurids were bipedal monsters, each back leg ending in three toes all armed with a powerful claw. They had well-muscled arms that may have been used to lash out at large browsing dinosaurs. The clawed fingers could certainly have ripped terrible wounds in the side of any unsuspecting prey. The powerful, muscular neck would have hammered the bladelike fangs into the victim's body so they could then slice off great chunks of flesh.

The Dynamic Dinosaurs

Most of what we know about dinosaurs has been learned from fossils. By studying the bumps, grooves, and scars on fossil bones, scientists have learned about the size and shape of dinosaur muscles. This helps determine body shapes and sizes. Some dinosaur bones show signs of injury, arthritis, cancer, and attack by other, more powerful dinosaurs. Fossilized skulls even tell us how big dinosaur brains were.

Fossilized eggs help scientists figure out how baby dinosaurs developed, how fast they grew, and what life was really like in a dinosaur "nursery." Fossil footprints or pathways tell us something about the social life of dinosaurs.

Making guesses

No matter how hard scientists look at dinosaur fossils, there are some things they can never be sure about. For example, what color were dinosaurs? Did dinosaurs wear camouflage markings? Perhaps forest species were dappled like fallow deer. Did species living in open country carry zebra-style black and white stripes to make it difficult for predators to see them? Meat eaters may have been spotted, like a cheetah, or striped, like a tiger. Perhaps it was only the big, herbivorous sauropods that were as gray as an elephant! We do not even know whether dinosaurs had hair.

Reading the footprints

Evidence about the life-style of the sauropod dinosaurs suggests that they may have given birth to live young, as mammals do. Sauropod pathways show that these giants moved in large herds as they searched for food. In order for a newborn sauropod to survive in a migrating herd, it would have to be able to get on its feet and start walking within a few minutes of hatching. A hatchling from even the biggest egg would not be able to keep up with the adults. A newborn dog-sized baby probably could. In any case, in a restless herd female dinosaurs would not have had time to stop and lay eggs, and then incubate them. The herd would have been hundreds of miles away by the time the hatchlings were ready to travel. So did baby sauropods keep up with their parents? The proof of the plodding lies in the pathways. Baby footprints appear quite clearly among those of the adults.

How fast did dinosaurs travel?

The speed at which an animal moves depends on its weight, the length of its back legs, and the length of its stride. Long legs and long stride help an animal to move quickly. Pathways and fossilized skeletons provide dinosaur detectives with all the data they need. The shape of the footprints even identifies the type of dinosaur. The "ostrich" dinosaurs were probably the fastest sprinters, reaching speeds greater than 30 miles per hour. Large dinosaurs like the 35-ton *Apatosaurus* (which used to be called *Brontosaurus*) may have been able to trot as easily as a 5-ton elephant. The massive 70-ton *Brachiosaurus*, on the other

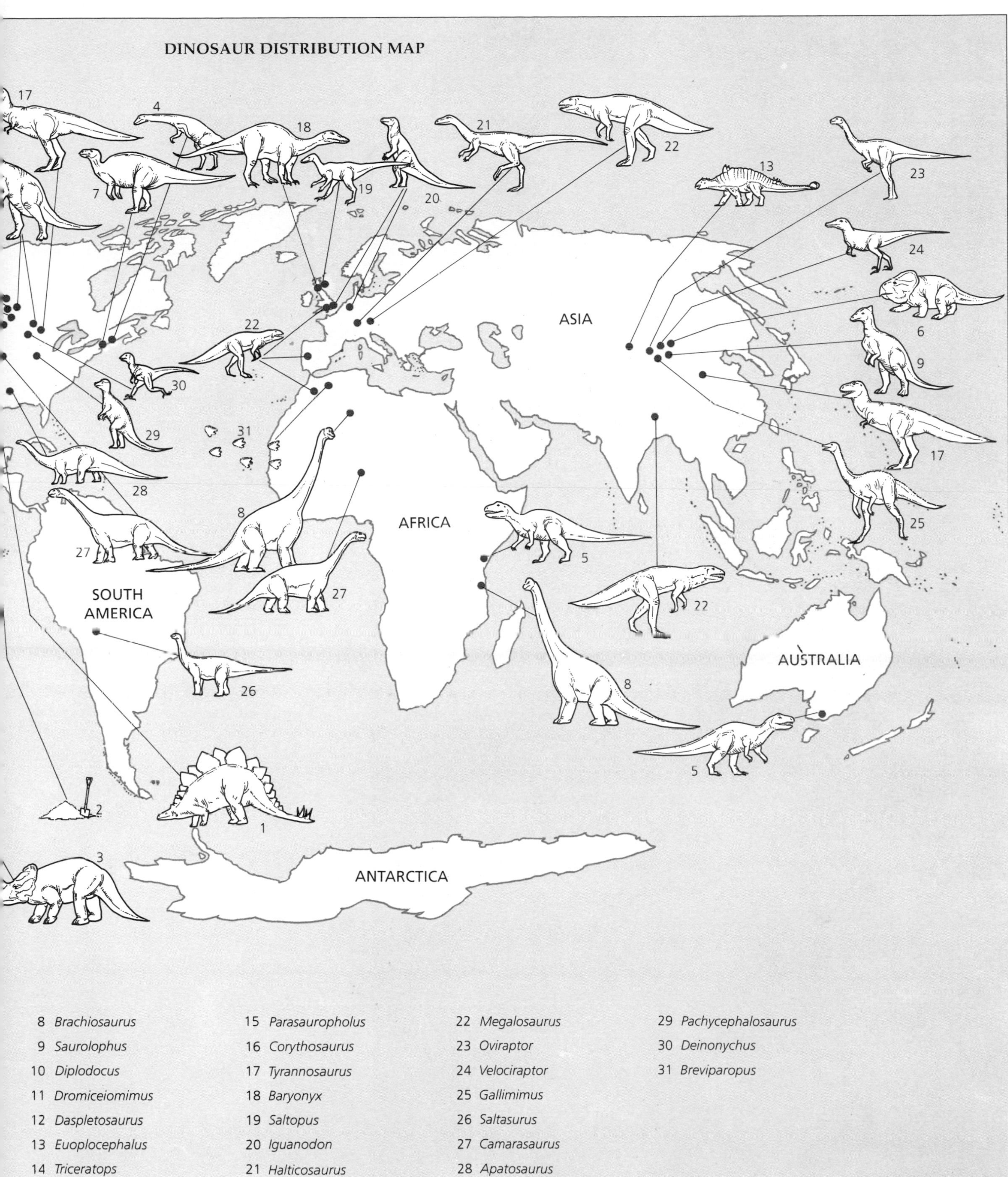
DINOSAUR DISTRIBUTION MAP
ASIA
AFRICA
SOUTH AMERICA
AUSTRALIA
ANTARCTICA
8 Brachiosaurus
9 Saurolophus
10 Diplodocus
11 Dromiceiomimus
12 Daspletosaurus
13 Euoplocephalus
14 Triceratops
15 Parasauropholus
16 Corythosaurus
17 Tyrannosaurus
18 Baryonyx
19 Saltopus
20 Iguanodon
21 Halticosaurus
22 Megalosaurus
23 Oviraptor
24 Velociraptor
25 Gallimimus
26 Saltasurus
27 Camarasaurus
28 Apatosaurus
29 Pachycephalosaurus
30 Deinonychus
31 Breviparopus

hand, probably managed only a slow walk at a steady 3-4 miles per hour.

The "thunder lizards"

Some of the sauropod dinosaurs were the biggest animals that have ever walked on earth. At 75 feet, *Brachiosaurus* (the "arm lizard") was as long as a tennis court, topping the scales at 75 tons. Its head was 40 feet above the ground, as high as a four-story building. *Supersaurus*, first discovered in 1972, has an estimated length of over 82-98 feet. Scientists are also still digging up *Ultrasaurus*, first discovered in 1979. At an estimated 130 tons in weight, *Ultrasaurus* may well prove to be the heaviest animal that has ever lived. But *Breviparopus* from Morocco may have been longer. So far only its tracks have been found. If its estimated length of 157 feet proves accurate, it was probably the longest backboned animal of all time.

A blueprint for size

The sauropod skeleton contained all the design features to cope with being big. The

◀ Jim Jensen, an American paleontologist, lying beside the shoulder blade of *Supersaurus*. The latest technology is now used to find more giant dinosaurs. Radar helped find *Seismosaurus* ("Earthshaker lizard") 13 feet under the rocks of New Mexico. No one yet knows its full size, but the signs are that it was at least 131 feet long.

▼► *Brachiosaurus's* backbone was designed like a bridge that spanned the front and back legs. The arched shape supported the very heavy belly. The long neck was built like a gigantic crane. The neck bones themselves formed the arm or "jib" of the crane. Bony ribs on the vertebrae worked like cables, while muscles and ligaments moved the neck from side to side and up and down.

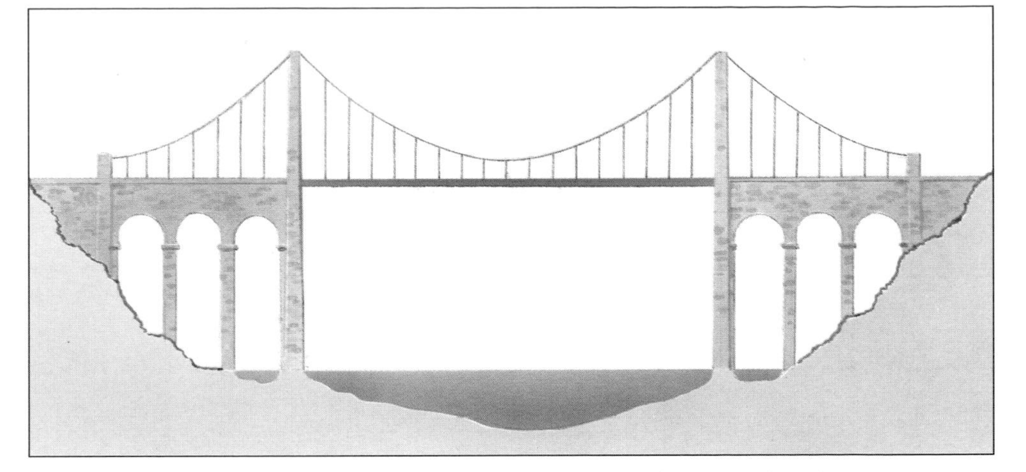

legs were massive pillars for supporting a huge weight. The foot bones were arranged vertically, rather than lying flat. They were also tied together with tough ligaments to give them greater strength. The upper part of a sauropod skeleton was designed differently. Here much less material was used and bones were often hollow. The skull and backbone were much more lightweight. Sauropods were "bottom-heavy" animals, another adaptation for the "big life." Some sauropods had a head smaller than a horse's and their brain was proportionally small. In some species it was no bigger than a kitten's.

The trouble with flat feet

Humans have a real problem with walking. Our toes lie flat on the ground when we stand, and the joint between the ankle and the lower part of the leg forms a right angle. This means that every time we take a step forward, we have to lift each heel off the ground in turn so that our body weight is transferred onto the ball of the foot. Try analyzing your walking movements. You move forward in a kind of bobbing action that causes your body to go up and down with each stride. Lifting the body up and down can be quite tiring, although it is not too bad for us because we are relatively lightweight animals and, anyway, shoes help. Did you know that shoes with heels (not too high) are actually more comfortable for walking in than flat ones and that they make walking easier and less energy sapping? This is because you are not spending energy on lifting your body up and down.

"Well-heeled" dinosaurs

For big animals, walking can be a major problem. Heavyweight animals such as elephants need large flat feet to support their enormous mass. But they would tire very quickly if the bones in their feet were arranged like those in a human foot, so that they rested the sole completely on the ground between strides and then lifted themselves up on the ball of the foot when they took a step forward. They would find rising up and down on their toes very hard going and would soon exhaust themselves. So each elephant's foot has a built-in heel made of tough fibrous tissue. They are positioned under the back of each foot and they keep the elephant on its toes while it is standing.

The great sauropod dinosaurs probably had a similar adaptation built into their feet. A 70-ton *Brachiosaurus* could not have bobbed up and down for very long but, then, it did not have to. Its broad toes were supported from behind by a thick wedge of tissue that functioned like a heel, just as in an elephant's foot. These specialized feet allowed the giant sauropod dinosaurs to plod forward using the minimum amount of energy and effort.

Did the giant sauropods live on land or in water?

For many years scientists thought that huge sauropods such as *Brachiosaurus* were amphibious. It was argued that they were too big and heavy to walk on land and that they could not have supported their great weight. Instead, scientists reasoned, they must have been aquatic (lived in water) and that their enormous weight was buoyed up by the water around them. It was also suggested that the long neck, together with nostrils positioned on top of the head, were an adaptation to living in deep water because they could function as a snorkel, thus allowing the animal to breathe when standing underwater. But we now know that this could not have been the case. The body of *Brachiosaurus* would have been at a depth of about 40 feet when completely submerged with just the top of its head sticking up above the surface of the water. At this depth, the water pressure would have been so great that the animal's lungs would have collapsed, so it would have been unable to breathe. In addition to this, the animal's high blood pressure would have caused heart failure. In any case, sauropods such as *Brachiosaurus* have all the characteristics of land animals. Their columnlike legs, deep and relatively narrow ribcage, and specially strengthened back are all features of land dwellers.

Hollow backbones for lighter bodies

Another piece of evidence to support the view that the great sauropod dinosaurs were terrestrial (lived on land) rather than aquatic is found in the remarkable adaptations some of them developed to reduce body weight. For example, the vertebrae in the backbone of sauropods such as *Apatosaurus* and *Diplodocus* had large cavities, called pleurocoels, in their sides. This hollowed-out design involved the minimum use of bone, and therefore helped reduce the animal's overall weight. If the sauropods were aquatic, there would have been no need for such adaptations for weight reduction because the water would have given their bodies all the support they needed, regardless of how heavy they were.

Dinosaur dilemmas

How do scientists, faced with the challenge of explaining aspects of dinosaurs' lifestyles that are almost impossible to calculate simply from studying fossils, figure out answers to questions like these:

What kinds of sounds did they make?

Were they good parents?

Did they sleep standing up?

What was their life span?

They might begin by comparing dinosaurs with living animals and asking more questions:

If plant-eating dinosaurs lived in herds like elephants and antelope, how were the babies looked after?

Did herds have a special system for keeping a lookout like herds of today?

Did they communicate with sound signals like geese or color signals like parrots?

Did carnivorous dinosaurs hunt in groups like lions or did they hunt alone like tigers or leopards?

By deducing answers to questions like these, often by a process of elimination, new ideas have been developed about dinosaurs' behavior and how they lived.

▶ The pathway in these two photographs shows a record of a dinosaur stampede. It took place at Lark Quarry in Australia and involved about 160 small dinosaurs. The herd probably panicked at the sight of a large predator. Can you see two different types of footprints?

Were dinosaurs warm-blooded?

The question of whether dinosaurs were warm-blooded is important because if they were then they were far more biologically "advanced" than scientists first thought.

Scientists have some evidence to support the idea that some dinosaurs may have been warm-blooded like mammals and birds: It is known that they walked on legs positioned directly underneath the body like mammals and birds; some had large blood vessels in their legs like mammals; long-necked types must have had a high blood pressure to get blood up to the brain – think of a giraffe.

An egg can be only so big. A fully grown *Apatosaurus* would have been about 100,000 times bigger than a baby hatching from the largest possible egg. Many scientists think that no animal could grow fast enough to increase its size by this amount. Fossils show that the leg bones of a young *Apatosaurus* were about one-quarter the size of an adult. Such a huge baby was probably too big to fit into even the biggest egg. This means it was probably born alive like a baby mammal. So some

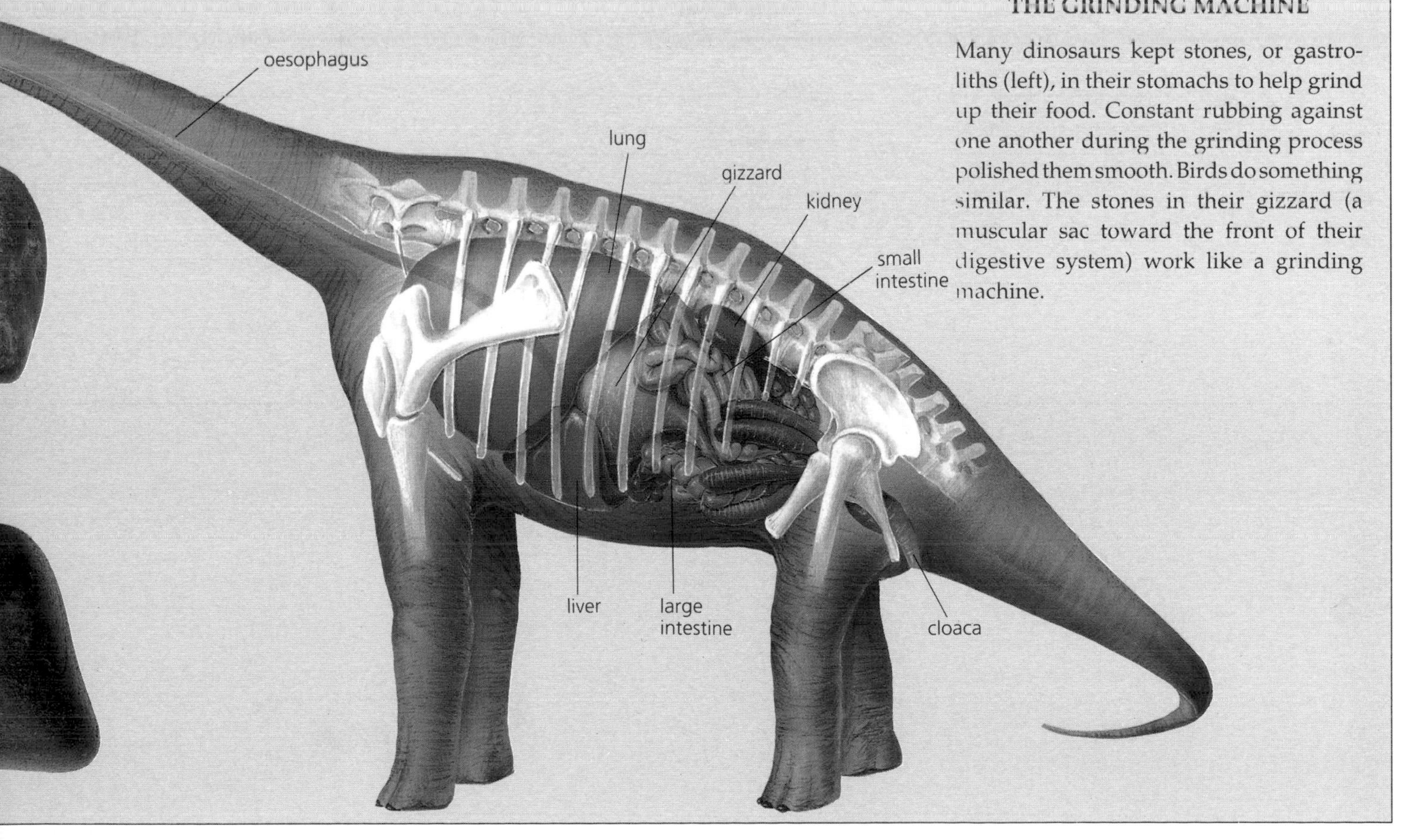

THE GRINDING MACHINE

Many dinosaurs kept stones, or gastroliths (left), in their stomachs to help grind up their food. Constant rubbing against one another during the grinding process polished them smooth. Birds do something similar. The stones in their gizzard (a muscular sac toward the front of their digestive system) work like a grinding machine.

scientists argue this way:" If you are going to be really gigantic you need to be born big. And you can only be born big if you are born living."

Appetite gives a clue to being warm-blooded. A lion eats its own weight in food about once a week. A cold-blooded Komodo dragon (the world's biggest lizard) takes about two months to match its weight with food. If the giant meat-eating dinosaurs were warm-blooded, their appetites must have been enormous. *Tyrannosaurus* would have needed to eat about a ton of food every day. Only a highly active, warm-blooded animal could catch this much food. But if *Tyrannosaurus* was cold-blooded, like the Komodo dragon, it could have survived on much less food. Which do you think it was?

▶ Cold-blooded animals grow more slowly than warm-blooded ones. A baby Nile crocodile is about 12 inches in length when it hatches. Within one year, if there is plenty of food available, it can grow to a length of 3 feet. A baby ostrich, which is warm-blooded, grows about twice as fast. When it hatches from the egg, it also measures about 12 inches from head to foot. In the first year of its life, it shoots up at the rate of about 6 inches a month and is fully grown within one year when it stands at more than 7 feet. A baby blue whale grows even more quickly. At birth it measures about 10 feet in length and weighs about 2.5 tons. When it is weaned, at about seven months, it has grown to more than 50 feet and tips the scales at about 23 tons, having gained nearly 220 pounds every day. It is not surprising that a baby blue whale grows so quickly, as it drinks the equivalent of about 2,500 glasses of milk every day! So how fast did dinosaurs grow? American scientists are now studying fossilized nests of hadrosaur dinosaurs to try and answer this question. Their findings may help to decide if some dinosaurs were warm-blooded.

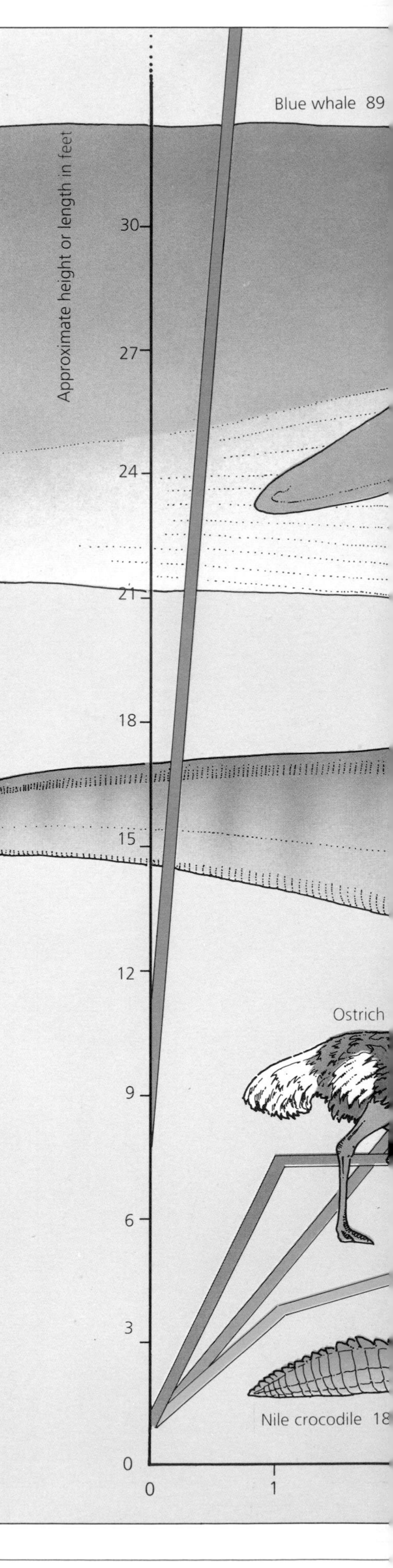

▼▶ The Komodo dragon (right), a cold-blooded reptile, eats about six times its own weight in food a year. The annual total for a lion (below) is roughly 50 times its own weight. The lion needs this enormous supply of food in order to keep up its high metabolic rate and maintain its warm-blooded condition.

?
Lambeosaurus 50 feet
3
4
5
6
7
8
9
10
11
12
Age in years

THE DINOSAUR DISCOVERERS

About 100 years ago, two Americans discovered a whole new world of dinosaurs. Othniel Charles Marsh (right) and Edward Drinker Cope made some of the greatest dinosaur discoveries of all time. They each hired teams of men to go west to Colorado and dig for dinosaurs. It was a hard life, and a risky one. The fossil hunters often faced attack from Native Americans who felt they were trespassing on their homelands.

Marsh and Cope became very jealous of each other, and tried to keep their finds secret. It is rumored that bitter quarrels broke out between the two, and that their teams even tried to steal each other's prize finds. It seems that "dinosaur rustling," like cattle rustling, was not unknown in the Old West. Hiring the necessary teams of men was a very expensive business, and Marsh and Cope each spent huge sums of money in trying to outdo the other.

Before the dinosaur rush to the Old West, only nine species of dinosaurs had been found in North America. In their lifetimes, Marsh and Cope discovered 136 new ones. Marsh had the edge: The score was Marsh 80, Cope 56. Marsh's team was responsible for finding *Allosaurus*, *Diplodocus*, *Stegosaurus*, and *Triceratops*. Cope's finds included *Camarasaurus*, *Monoclonius*, and *Coelophysis*.

Roy Chapman Andrews led the first dinosaur expedition to Mongolia in 1922. He and his team used cars and camels to reach remote dinosaur country. Their most famous dinosaur finds were the fossilized bones, eggs, and nests of *Proceratops*. Since these early explorations, many new and exciting discoveries have been made. These include the world's richest find of late Cretaceous dinosaurs: thousands drowned in prehistoric floods 70 million years ago.

Robert Bakker, an American scientist, is one of the more modern "thinkers" about dinosaurs. He has produced some startling theories. His world of dinosaurs is very different from the one previously imagined. Bakker's dinosaurs are energetic, light-footed animals. The big sauropods were herd animals and the meat eaters were cunning hunters. Perhaps most important of all, Bakker thinks dinosaurs might have been warm-blooded.

Jack Horner is sometimes called "the man who walks on eggshells." He has made many exciting discoveries, including whole nesting sites of hadrosaurs (duck-billed dinosaurs), their fossilized eggs, and babies of all ages. He discovered these in 1978 with another scientist, Robert Makela. Perhaps his research into a dinosaur's nursery days and how fast dinosaurs grew will help solve the argument over whether or not dinosaurs were warm-blooded.

THE CRETACEOUS PERIOD

144 MILLION TO 65 MILLION YEARS AGO

During the Cretaceous, the world's great land "breakup" continued. The enormous landmasses of Laurasia and Gondwanaland continued separating. The Atlantic Ocean became wider as South America and Africa pulled apart. Africa, India, and Australia also began to move away from one another to form gigantic "islands" south of the equator. Most of what we now call Europe was underwater.

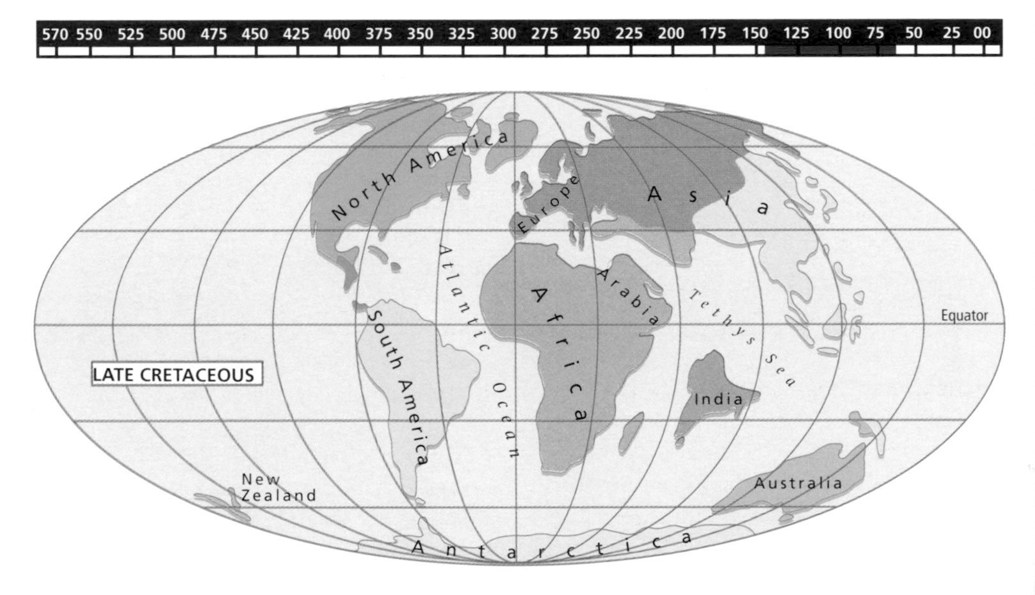

Forces within the earth continued to shape its appearance on the surface. Much of the earth's surface was covered by warm, shallow seas. At the beginning of the period, many new forms of mollusks evolved and diversified, such as the mussel-like bivalves and snail-like gastropods. The great molluscan ammonites were less common, but the belemnites were important marine dwellers. New species of meat-eating crustaceans such as shrimps, crabs, and lobsters also appeared in the shallow waters close to land.

Fast-swimming predatory reptiles were common farther out at sea. Some of the Jurassic plesiosaur and ichthyosaur species still survived. But later, ferocious sea-going plesiosaurs such as *Elasmosaurus* and the lizardlike mosasaurs also came on the scene. They fed on newly evolved bony fish, and cartilaginous species such as skates and rays. This was the time when the gigantic turtle, *Archelon*, paddled through the warm waters, hunting for food. *Archelon* measured nearly 13 feet in length.

Snakes, birds, and bees

On land, many exciting things were happening. Snakes had now evolved from lizardlike ancestors. There were also some new seabirds. *Hesperornis* looked like a large diver. *Ichthyornis* was a small ternlike bird, and the first to have a real breastbone. Bees, moths, and other insects were everywhere, and spiders lurked among leaves waiting to catch them.

An "explosion" of flowers

Up until the beginning of the Cretaceous, distribution of spores and pollen was a risky business. Many primitive plants had to rely on wind to do the job. Even worse, other plants had no dispersal mechanism at all. They simply dropped their spores on the ground where they grew. But in the Cretaceous new and more efficient ways of spreading pollen developed. Flowering plants (angiosperms) were now evolving in partnership with the insects.

A special relationship was beginning that has lasted to this day. This partnership quickly began to improve the chances of successful pollination. Insects provided an economical service for the collection and delivery of pollen. But the flowering plants had to "pay" for this by offering "bribes" of something worth collecting. Brightly colored petals and sweet scents acted as the temptation. Delicious, sugary nectar and stores of pollen were the food source the insects needed. The flowering plants began to spread rapidly over the earth's surface. Today there are more than 250,000 different species of flowering

▲ Sometimes, even soft parts of living things like flowers can be fossilized. This is a fossil flower bud. It was found in Cretaceous mud in Sweden. When alive, it was only 1/12 of an inch long and 1/25 of an inch across.

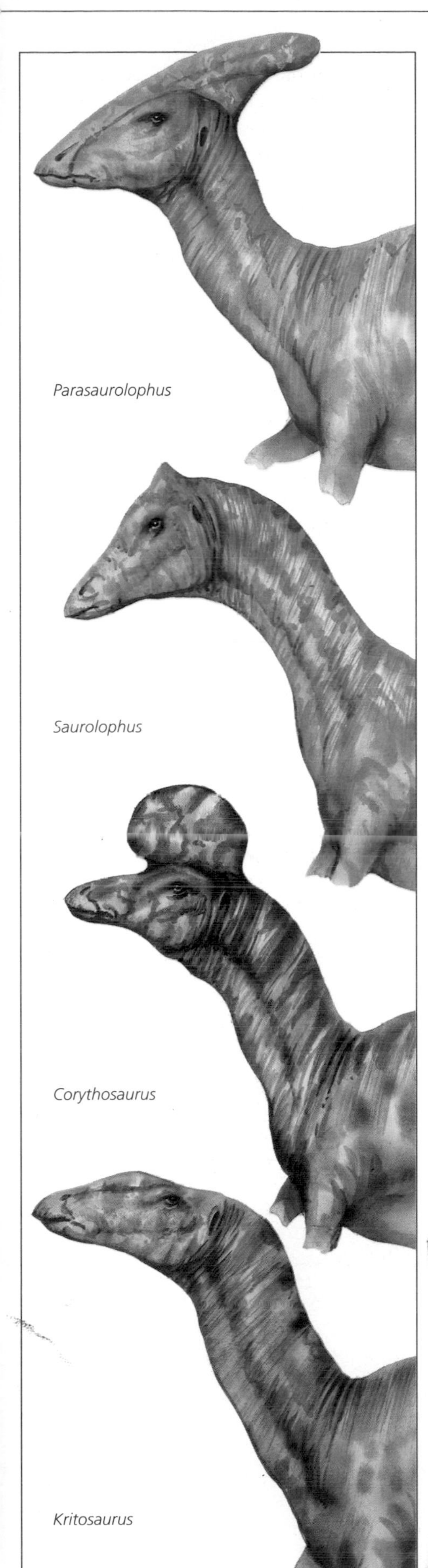

◀ The crests of hadrosaurs probably also functioned as honking devices, or sound resonators. Each species had its own specially shaped "trumpet," producing a characteristic honk, squeak, grunt, or bellow that could be recognized as a sound signal by its own type.

plants compared to 50,000 species of all other green plants put together.

New dinosaurs for new plants

New types of plants sprang up over the land and huge forests began to appear. Land animals found different kinds of leaves to eat and vegetation to chew. Dinosaurs were still evolving and new species appeared throughout the Cretaceous. The duck-billed hadrosaurs were the dominant grazers of the period. They lived in herds like today's antelope. Their mouths were full of teeth, sometimes as many as 2,000 in a single skull. They were the perfect machinery for chopping and grinding tough plant material.

Other plant eaters also became more common. New stegosaurs and ankylosaurs appeared. Toward the end of the Cretaceous, *Triceratops*, one of the last dinosaurs to evolve and the biggest of the ceratopsians, bulldozed its way across the North American landscape.

Thick heads, crests, frills, and flanges

In the Cretaceous, crests, frills, and flanges became even more elaborate than they had been in the Jurassic. Hadrosaurs wore a variety of head crests. For a long time, scientists thought these dinosaurs were aquatic and that these crests worked as aqualungs or snorkels. A recent theory is that these helmets acted as identification tags. They were a kind of head badge or signal to help males and females of the same species recognize each other.

Ceratopsian dinosaurs had strange horns and huge, backward-pointing neck frills. These gave protection from both head-on attacks and danger from behind. They may also have helped in keeping the animals cool by acting as heat and light deflectors. *Ankylosaurus* was built like a small army tank and covered from head to tail in hard, bony plates. It even had bony eyelids. Its tail ended in a large, bony lump that it probably used like a club. Such armor plating probably made *Ankylosaurus* a match for even the fiercest predator.

Pachycephalosaurus

Male *Pachycephalosaurus* dinosaurs could have been head butters supreme. The skull was made of massive bone up to 10 inches thick, which may have doubled up as a battering ram and a crash helmet during territory fights and mating disputes over females. If such fights took place, this extra protection would have stopped two males from bashing each other's brains out. Contestants may have collided with such force that the deceleration on each head would have been nearly 20 *g*. This is more than twice the gravitational force experienced by a pilot making a tight turn in an F16 jet.

▼ Two male *Pachycephalosaurus* engaged in a fight for dominance. Some scientists believe this is how *Pachycephalosaurus* ("dome-headed" lizard) fought for females and territory.

The tyrannosaurids

The tyrannosaurids were among the biggest flesh-eating animals that have ever lived and their fossil remains have been found in North America and central Asia. All tyrannosaurids had a massive head and a huge body that was carried on large, muscular back legs, each ending in three toes. They also had a long, powerful tail that probably acted as a counterbalance when they walked upright.

Tyrannosaurids had surprisingly small front legs for such huge animals. One suggested function is that they were used as props to help an animal get up from lying on its stomach after resting.

Ferocious hunters?

Scientists are undecided about some aspects of tyrannosaurid life-style. Were they really ferocious predators that caught their prey by sprinting after it, or were they scavengers feeding on the dead bodies of other dinosaurs? The shape of the skeleton, with its long, powerful tail counterbalancing the trunk, suggests they were fast-running animals over short distances.

Rather than chasing their prey, perhaps tyrannosaurids lurked in ambush and then pounced. The shape and size of the head supports this idea. The skull is made of heavily reinforced bone. This suggests that it had to withstand an enormous impact when a tyrannosaurid ran into its prey with its jaws wide open.

Some experts believe tyrannosaurids were too big to move quickly and, instead, relied on an acute sense of smell to help them find rotting corpses on which to feed. These scientists also argue that the tyrannosaurids' long teeth would have shattered if they had been driven with force into the body of another large dinosaur. Their theory is that the teeth were used as a set of steak knives to slice off large lumps of meat from an animal that was already dead.

The tyrant lizard

Tyrannosaurus rex is probably the best known of all dinosaurs and was probably also one of the last to evolve. It was a massive animal measuring 40 feet in length and standing as tall as a two-story building. Its gigantic skull was more than 3 feet long and was armed with rows of formidable-looking teeth, each about 6 inches in length.

In 1990, a team of paleontologists led by Dr. Jack Horner dug out a nearly complete *Tyrannosaurus rex*. This find has helped scientists to rethink their ideas about this most famous of dinosaurs. For example, it has been reduced in size – it is now thought to have been only a 4-ton monster with the intelligence of an emu (not very bright!). By studying its coprolites (fossilized droppings) and also the chemical makeup in its bones, experts think it was probably both hunter and scavenger.

Dinosaur DNA

In 1993 scientists working in America found some fossilized blood from *Tyrannosaurus rex*. Unlike mammalian blood, reptilian blood contains red blood cells that have nuclei. This means that scientists will now be able to study *Tyrannosaurus*'s genetic makeup and even the structure of some of its genes. When the experts have completed their work, we will know far more about its life history, how its body chemistry worked, how it lived, and how it behaved.

▼ If *Tyrannosaurus* was one of the biggest carnivores, *Deinosuchus* was certainly the largest-ever crocodile. It was 40 feet in length with massive jaws. It lurked in rivers and fed on animals coming to drink.

▲ This African crocodile has been drawn to the same scale as *Deinosuchus*. You can see how much smaller it is.

TYRANNOSAURUS'S TOOTH

This *Tyrannosaurus* tooth is actual size. You can see why some scientists think this meat-eating dinosaur was a scavenger. The zigzag edges along the blade make each tooth a perfect meat slicer.

► *Tyrannosaurus*: one of the biggest carnivores that has ever lived. The first virtually complete *Tyrannosaurus* skeleton was found in Montana in 1902.

Dragons of the air

Pterosaurs still patrolled the Cretaceous skies. They had evolved a long way since the Triassic, and many new forms now existed. *Pteranodon* was a giant, tailless, ocean-going fish eater. It was one of the most advanced of the pterosaurs. It probably used its 26-foot wingspan to glide over the surface of the oceans. Some of the fish caught at sea may have been stored in a throat pouch for feeding to its young. Perhaps it even swallowed and partly digested some of the catch before regurgitating it back at the "nest."

Its head crest may have acted like a weather vane to keep it pointing headfirst into the wind. Or it may have had another function. *Pteranodon's* massive beak must have been a problem. A sudden gust of wind catching it would have twisted the whole head around so violently that its neck would have certainly snapped. To prevent this, *Pteranodon's* equally big head crest would have resisted twisting, and also acted as a counterbalance and rudder.

From slipped disks to flatulence

Something very odd happened at the end of the Cretaceous period. The belemnites, ammonites, pliosaurs, plesiosaurs, ichthyosaurs, and mosasaurs all became extinct from the world's oceans. The dinosaurs suddenly died out on land, and the pterosaurs disappeared from the Cretaceous airspace. Scientists have been

▲ Nearly 65 million years after it disappeared, *Quetzalcoatlus* flew again when Dr. Paul Macready and his team of engineers built and flew a half-sized model. The last animals to see such a sight were the dinosaurs. The model had a wingspan of 18 feet and weighed 44 pounds. It carried a radio receiver, an automatic pilot system, sensors, 56 batteries, and two electric motors to flap the wings. The model flew successfully on many occasions in Death Valley in California. Sadly it crashed on its first public appearance at Andrews Air Force Base in California, on May 17, 1986.

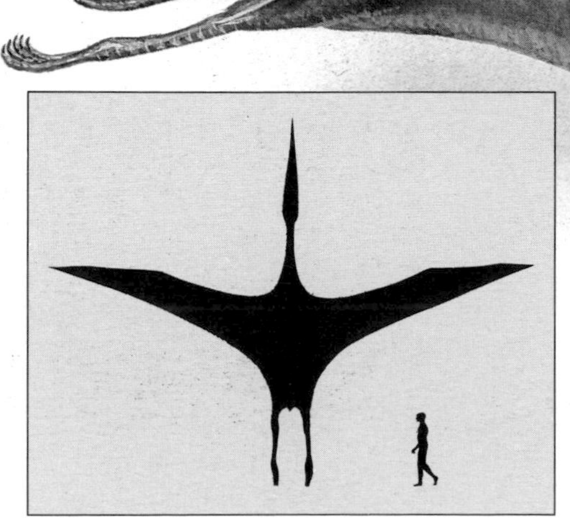

◄▲ *Quetzalcoatlus*, or "feathered serpent," was a long-necked pterosaur that soared on hot air rising from the ground, and flapped its way through the skies above Texas and Alberta, Canada, in the late Cretaceous. It weighed over 220 pounds and its outstretched wings were 40 feet across – about as big as a Spitfire jet's.

puzzling for a long time about this mass extinction and especially why the dinosaurs suddenly died out.

Many theories have been put forward. Some scientists suggest that the earth's climate changed as temperatures became cooler. Others support the theory that a gigantic asteroid collided with Earth. Such a collision would have caused a huge cloud of dust that would have blotted out the Sun, on which all living things depend. Maybe dinosaur eggs were eaten by the up-and-coming mammals. Perhaps the biggest dinosaurs suffered from slipped disks because of their great weight. Other theories lay the blame on a sudden enormous outbreak of volcanic activity, or a sudden massive downpour of acid rain following Earth's collision with the asteroid. A scientist from Indiana has even suggested the startling theory that the dinosaurs killed themselves off with their own flatulence (wind). The methane they produced caused the earth's atmospheric temperature to warm up, creating a kind of "greenhouse effect." This might have increased global warming to a point where dinosaurs could not stand the heat.

Rise of the mammals

Early mammal-like animals first appeared in the Triassic a little over 200 million years ago. They were shy little animals that spent most of their time scurrying around looking for insects on which to feed. In evolutionary terms, they did not do much for nearly 100 million years. Then, in the first half of the Cretaceous period, evolutionary changes started to take place among these secretive early mammals. These changes resulted in the development of the monotremes, the marsupials, and the placental mammals. These were the animal types that were set to take over from the dinosaurs at the end of the Cretaceous, 65 million years ago. They have been the dominant form of life on earth ever since then.

THE ASTEROID THEORY

The rocks formed at the end of the Cretaceous period show evidence that supports the asteroid theory about the death of the dinosaurs. They contain a thin line of iridium, a rare element on Earth but one common in asteroids. Such an asteroid must have been enormous – at least 6 miles in diameter with a weight of at least 4 million tons. But a huge object like this crashing into Earth would have left a crater scar at least 62 miles across. Until recently, no such telltale crater had ever been discovered. Now the missing evidence may have come to light. In 1992 Dr. David Krill and Dr. William Boynton from the University of Arizona discovered a gigantic underground crater in the Yucatán Peninsula, in Mexico. It measures 112 miles in diameter, and dates back to the exact time when the dinosaurs disappeared 65 million years ago. Unlike a detective story, this may not be the "smoking gun" that scientists have been seeking for the last 10 years, but it may be the "bullet hole." This huge crater may be the final answer scientists have been looking for to solve the disappearing dinosaur puzzle.

A Dinosaur Spotter's Guide

Imagine someone has invented a time machine that allows you to travel back through the ages. You are interested in dinosaurs and decide to use the machine to go back to the late Cretaceous period. You are anxious to photograph and identify as many dinosaurs as you can during your visit to your chosen time warp. You feel confident that you will be able to recognize the more common species, but feel less sure about many of the other dinosaurs you are likely to meet. In order to get over your problem, you ask a dinosaur expert to help you.

The expert decides to make up an identification key to help you in your task. Remember an identification key is designed to help scientists recognize animals and plants that they do not know the name of or have never seen before. It works a little like a treasure hunt. It consists of a series of clues that are arranged in pairs. By answering each pair of clues in turn (starting at the beginning) the user is led on to another pair (not necessarily the next in order) until, in the end, they are able to put a name to the specimen they chose to identify.

Here you can see part of the expert's basic identification key that has been modified to make it easier to use. This shows you how a key works.

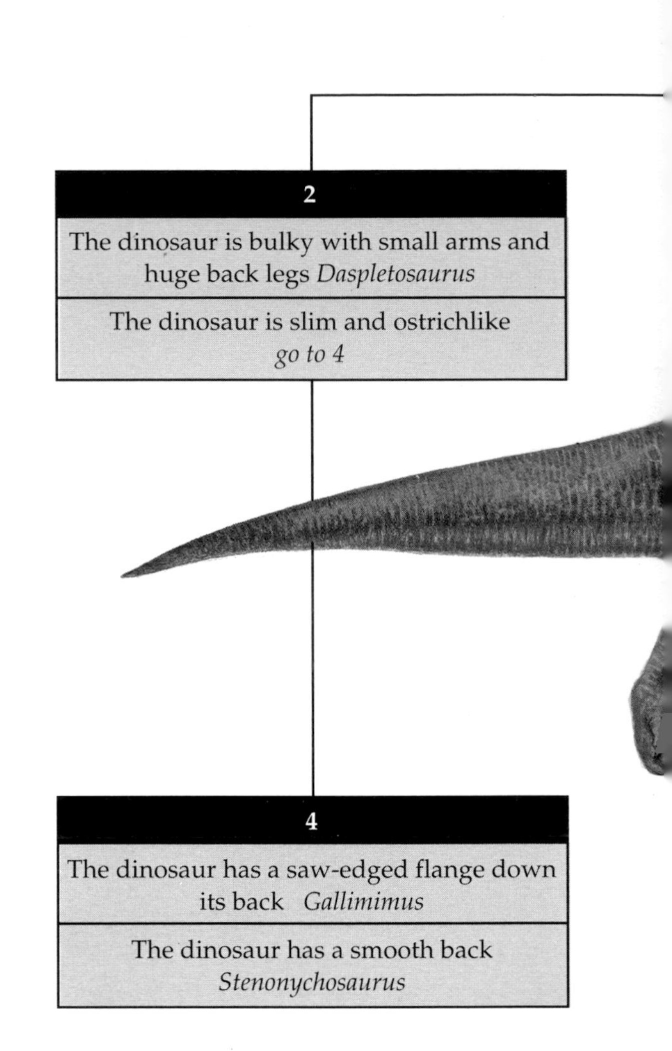

2
The dinosaur is bulky with small arms and huge back legs *Daspletosaurus*
The dinosaur is slim and ostrichlike *go to 4*

4
The dinosaur has a saw-edged flange down its back *Gallimimus*
The dinosaur has a smooth back *Stenonychosaurus*

DINOSAUR ?
1
The dinosaur is bipedal *go to 2*
The dinosaur is quadrupedal *go to 3*
3
The dinosaur has a bony neck shield *Triceratops*
The dinosaur does not have a bony neck shield *go to 5*
5
The tail ends in a bony ball *Euoplocephalus*
The tail ends in twin spikes *Scolosaurus*

The Paleocene Epoch

65 MILLION TO 55 MILLION YEARS AGO

The Paleocene was the beginning of the Cenozoic era. The continents were still on the move as the great southern continent of Gondwanaland continued to split up. South America was now completely cut adrift with its own unique floating "ark" of early mammals. Africa, India, and Australia moved even farther apart. Australia kept close to Antarctica throughout the Paleocene. Over much of the Earth's surface, more dry land was exposed as the sea level dropped.

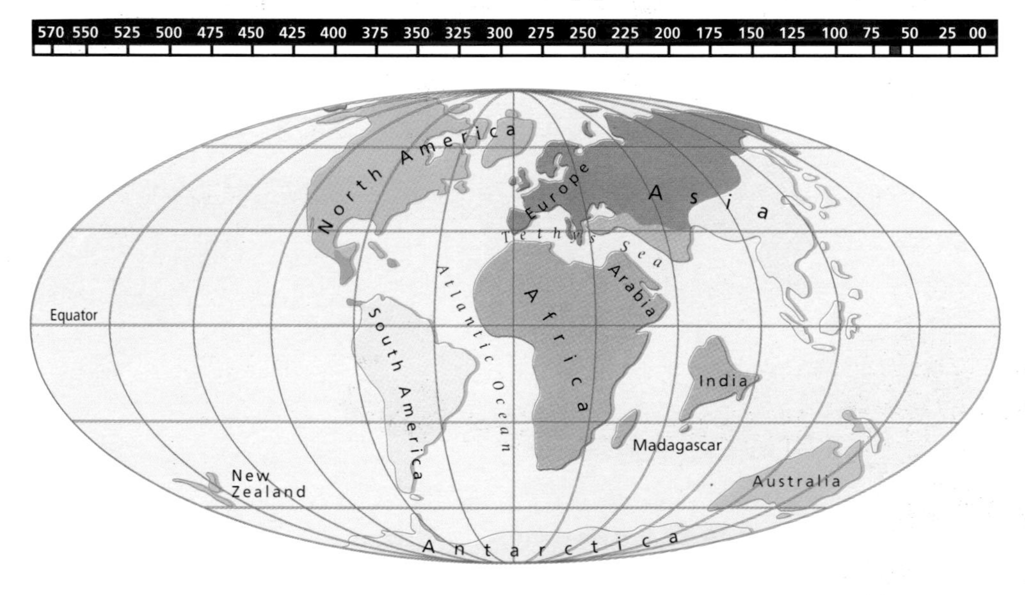

During the Paleocene, new species of gastropods and bivalves replaced the extinct ammonites as the main types of mollusks in the oceans. New forms of sea urchins and foraminifers evolved. Many gaps were left in the sea's food chains by the disappearance of the ichthyosaurs, the plesiosaurs, and the other marine life that became extinct at the end of the Cretaceous. New meat-eating bony fish and sharks replaced these extinct reptiles as the main oceanic carnivores.

A chance for the mammals

The world was about to enter the age of mammals. Three groups – the monotremes, the marsupials, and the placental mammals – took over from where the dinosaurs had left off. Mammal-like animals had already appeared toward the end of the Triassic, but, unable to compete with the then dominant dinosaurs, these mammalian prototypes had hidden away among their more successful competitors.

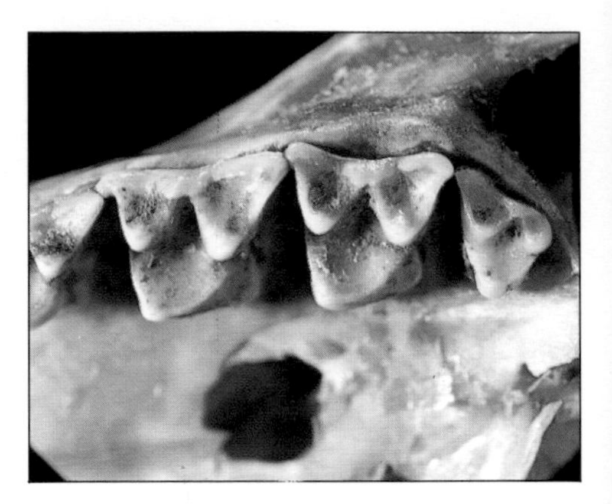

▲ A close-up of some of the teeth of a tree shrew, a primitive mammal alive today. The sharp, triangular shape of each tooth is typical of a mammal. The early mammals of the Paleocene probably had teeth like this. Scientists know about the first mammals mainly from their fossil teeth and bits of jawbone. From these it is easy to determine that these secretive little animals were nocturnal insect eaters.

THE MAIN DIFFERENCES BETWEEN REPTILES AND MAMMALS

Mammal
- covered in fur
- warm-blooded
- gives birth to live young
- feeds young on milk

Reptile
- body covered in dry scales
- cold-blooded
- lays leathery-skinned eggs

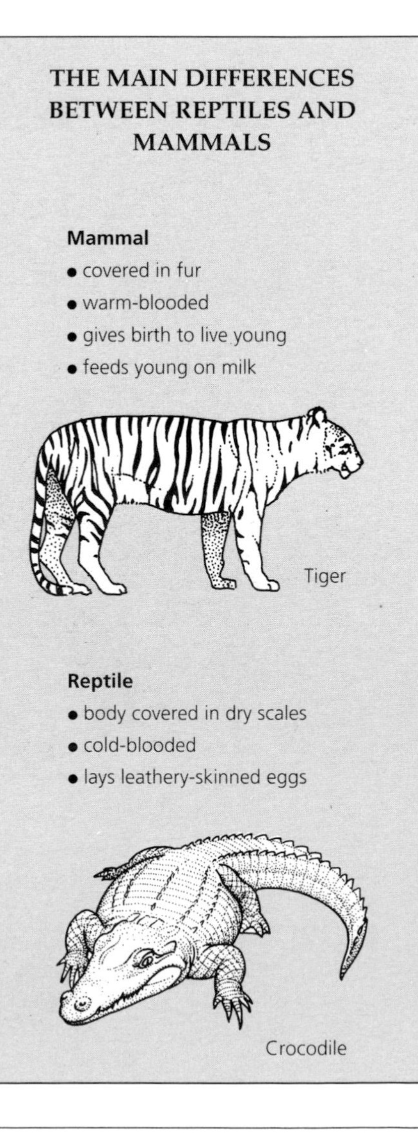

New food for new mammals

Some early mammals remained insect eaters. The first shrews and hedgehogs did this, competing with other ground feeders, such as frogs and toads. But there were plenty of insects other than those on the ground. Some mammals became airborne and began to feed on the flying insects that formed a huge pantry of untapped food. But the disappearance of the dinosaurs also meant there was now plenty of other food available as well. Many feeding niches had been left empty by the dinosaurs' sudden extinction. Other early mammals now started coming out in the day to eat a variety of diets. There were also rodentlike gnawing animals called multituberculates, and tree-living, squirrel-sized primates that fed on a mixed diet.

Early mammals quickly evolved into shapes and sizes that suited them for life in almost every habitat. The smallest were probably still insectivores. Bigger versions hunted or ate carrion (dead animals). Hefty herbivores and meat eaters appeared. Amblypod ("slow-footed") mammals were clumsy browsers (eaters) of leaves and other vegetables. Some had tusks and strange horns as defense weapons. An amblypod called *Barylambda* was the size of a small pony. It may have sat on its haunches to reach leaves higher up. Creodonts were flat-footed, meat-eating mammals. Some were as small as a weasel. Others were bigger than the largest bear.

The monotremes – the egg-laying mammals

The monotremes first appeared in the mid-Cretaceous. They are the most primitive group of living mammals. Only three species still survive from the days of the Paleocene: There are two species of echidna, or spiny anteater, and one species of platypus. All are found only in Australia and New Guinea. Although monotremes are mammals, they do still have one main reptilian characteristic: They lay eggs.

◀ Amblypods were early hoofed herbivores. One of the first was *Pantolambda*. It was the size of a sheep with heavy legs, short feet, and large canine teeth. It may have lived like a hippopotamus, wallowing in mud and browsing on land.

▲ The male platypus is unusual in being one of the very few poisonous mammals. It can inject enough poison through its two ankle spurs to kill a dog!

The platypus – a hairy "duck"

The monotreme platypus first became known to scientists in Europe in 1798 when a dried skin was sent to Britain from Australia. At first an unknown taxidermist (someone who preserves animals) was suspected of stitching a duck's beak onto the body of a mammal. There does not appear to be any record of what scientists thought about the strange tail! Experts argued over this curious animal. Eventually they agreed that what they were looking at was an unusual early type of mammal.

In the last 200 years, we have learned much more about the platypus and its way of life. For example, we know that the female lays her two eggs at the end of a special breeding burrow in the bank of the river where she feeds. The young hatch in a very underdeveloped state after about 10 days. Then they spend three or four months feeding on their mother's milk before entering the outside world. Although monotremes feed their young with milk, they do not have teats. We also know that the platypus's curious "duck's beak" is a super touch-sensitive bill. It is used for "feeling" for food on the river bottom. It is also receptive to electric fields, which it uses to detect prey.

Marsupials – the pouched mammals

The first marsupials date from the middle to late Cretaceous (about 100 million years ago) in North America. Later, in the Eocene, they reached all continents except Africa and Asia. They also moved across Antarctica to Australia. Marsupials are more advanced than the monotremes. Today, 266 species still survive. Most of them live in the Australasian region and the New World (mainly South America). Instead of laying eggs they give birth to live young. The bigger species have a pouch that acts like a built-in nursery, where each baby stays until it is big enough to look after itself. Most of the smaller marsupials are pouchless.

The kangaroo – a red giant

The red kangaroo is the biggest of all marsupials. At 7 feet in height and 176 pounds in weight, it is a giant by any standards. But a baby kangaroo, or "joey," starts life at birth weighing less than an ounce. However, as soon as it climbs into its mother's pouch and starts to suckle, it begins to grow quickly. It spends about seven months as a passenger, gradually taking time off to come out and explore the world outside. However, even a large baby will scurry back into its mother's pouch for safety when there is danger. It sometimes manages to do this while its mother is on the run.

Placental mammals

While the monotremes and marsupials were developing their own methods of reproduction, another group of mammals (the placentals) was beginning to develop a different way to produce offspring. This method centered around a structure called the placenta and involved keeping offspring inside the female's body until they were much more developed (unlike marsupials, whose young are born very underdeveloped). This new method of breeding obviously had great advantages. Because they were bornin a much more developed state, young placentals had a much better chance of survival. The placentals also perfected a method for suckling their newborn offspring and they developed new patterns of behavior that included long periods of parental care.

The best of both worlds

Birds have always been warm-blooded, but they have never improved on the egg as a way of reproducing. It may be that some dinosaurs could control their body temperature, and perhaps some abandoned laying eggs in favor of live birth, but we need to know much more about dinosaur behavior before we can be sure of this. However, for the moment it looks like the placental mammals were the first to combine the best of evolution's reproductive inventions: warm-bloodedness and the production of well-developed young. Having perfected this new breeding technique, placentals were able to explore more of the world and exploit a much greater range of habitats.

In the Paleocene, many placentals were small animals much like their Cretaceous ancestors. However, they soon began to compete with the marsupials. They evolved quickly and many new species appeared. Their ability to maintain a warm body temperature, combined with their improved reproductive behavior and large brain, helped them become a very "successful" group of animals, and they gradually spread to all parts of the world. They are now the most successful of all vertebrate groups. There are about 4,000 species alive today. They include all the well-known types, such as dogs, cats, bats, whales, monkeys, and apes. They live in almost every type of habitat. They have adapted to the hottest deserts and to the coldest regions on earth. They have also adapted to life in both air and water.

PLACENTAS AND POUCHES

Placental mammals are much more developed when they are born than either monotremes or marsupials. Even so, baby placentals still need lots of care and attention from their parents as they grow. For the early part of their life they also feed on milk produced by the female.

A newborn kangaroo, or joey, is 3/4 of an inch long and 1/30,000 of its mother's weight. As soon as it is born, it crawls into its mother's pouch using its well-developed front claws. The journey takes about 3 minutes. The mother suckles the baby in the pouch for about 7 months and for up to a year outside the pouch.

The Eocene Epoch

55 MILLION TO 38 MILLION YEARS AGO

In the Eocene the main landmasses were beginning to move to the positions they are in today. Much of the land was still divided into gigantic islands as the great continents separated even farther. South America lost links with Antarctica, and India moved closer to Asia. Antarctica and Australia were still close together at the start of the Eocene. Climates worldwide were warm or mild. Much of the earth's vegetation was lush and tropical with great areas of swampy forest.

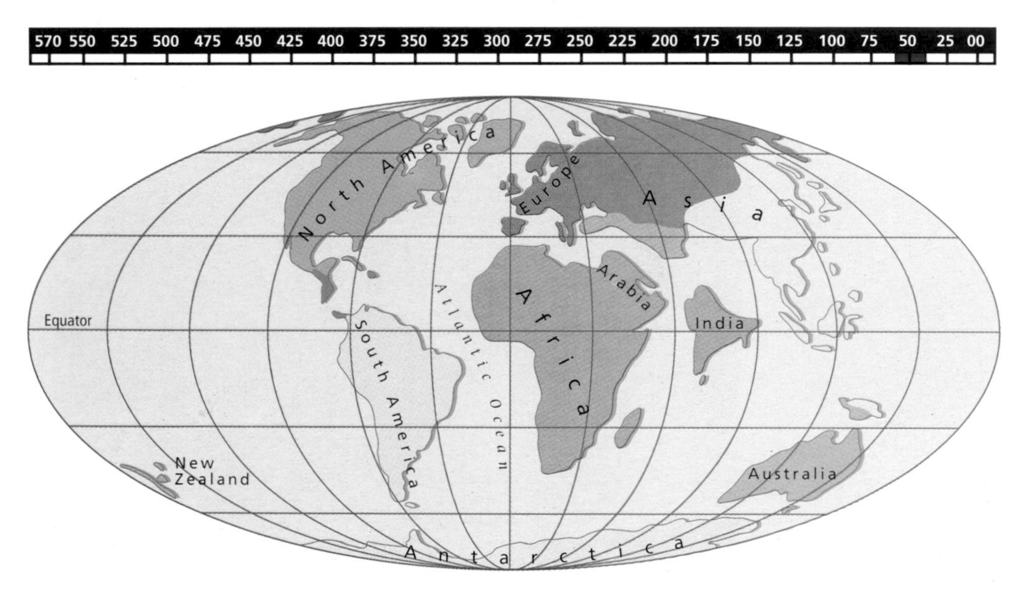

During the Eocene the oceans were rich in different kinds of floating plankton, made up of masses of tiny animals and algae. Beneath the surface there were new species of mollusks and also crustaceans such as crabs and crayfish. There were bony fish of all shapes and sizes, too. The fossil record also shows there were many new species of fish in freshwater lakes and rivers in many parts of the world.

Mammals galore

The plant-eating mammals of the Paleocene evolved into many new types in the Eocene. Early on, small, five-toed, hoofed animals called condylarths fed on soft plants and leaves. These fast-running animals shared the ancestry of modern horses, cattle, pigs, tapirs, rhinoceroses, and deer. There were also large herbivores such as *Coryphodon* and *Uintatherium*. Rodents were the main group of small mammals. The ancestors of today's lemurs, tarsiers, and galagos lived in the trees.

Return to the sea

Although meat-eating sharks and some carnivorous bony fish replaced the great marine reptiles at the end of the Cretaceous, some land mammals also returned to the sea to take the place of earlier ichthyosaurs and plesiosaurs. The mammals that later evolved into whales did this. The earliest fossil whales date from the Eocene. These aquatic mammals probably evolved from a group of flesh-eating hoofed mammals that returned to the water. Perhaps they did this to avoid competition with other animals on land.

Toward the end of the Eocene, creatures similar to today's whales lived in the oceans. Whales are mammals that are perfectly adapted to life in water. Their streamlined body, large front flippers, and powerful horizontal tail fins (flukes) make them strong swimmers. The fastest whales are capable of speeds up to 37 miles per hour. They never come onto land, even mating and giving birth at sea.

The problems of moving through water

Water is about 800 times denser than air. It pulls back (drags) on any animal trying to move through it. In order to reduce drag, marine animals need to be streamlined. In whales, blubber is a fatty substance that gives the body a smooth shape by filling in any hollows and smoothing out odd bumps on the skin. Further streamlining is achieved by reducing parts that stick out. A whale's back legs have disappeared and there are no external ears. The male's penis is completely hidden in muscular folds and the female's teats are tucked away in slits. All these adaptations give whales a super- streamlined shape. Something else helps whales swim quickly. It has now been discovered that they accelerate by "jumping out of their skins." As they move forward, they leave behind a very thin "skin ghost" of themselves. The water drags on this and the animal escapes, rather like the way a piece of slippery soap shoots out of your wet hand when you are in the bath.

Other newcomers

There were a number of other "newcomers" in the Eocene. The fossil record as it stands suggests that the first ants and bees evolved at this time. The first starlings and penguins also appeared and so did the earliest poisonous snakes.

EVOLUTION OF WHALES

Zeuglodon was a prehistoric whale. Its body was very flexible, a little like a snake's. It reached 66 feet in length and had saw-edged teeth for catching fish. Compare *Zeuglodon* (top) to the modern sperm whale (middle) and bowhead whale (bottom).

◀ Since the Eocene, seas have been rich in phytoplankton. These microscopic, "wandering" algae live in enormous numbers near the surface of the oceans today. Just 3 cubic feet of seawater may contain as many as 200,000 of them. They contain chlorophyll, so they can change the sun's light energy into chemical energy by photosynthesis. They form a kind of saltwater "soup" on which microscopic floating animals (zooplankton) feed. These, in turn, are fed on by bigger animals. This is how oceanic food chains are built up. The world's biggest animal today (the blue whale) eats a special kind of zooplankton called krill (left). It gulps down 10 tons of these tiny shrimplike animals in a single meal as it filter feeds its way through the sea.

THE FIRST HORSE REPORT

The first horse was a small, fox-sized animal (about 12 inches at the shoulder) that lived in the early Eocene. The oldest fossils were found in rocks in England in 1840. The animal was named *Hyracotherium*. It lived in the great swampy Eocene forests of North America and Europe, where it munched the leaves of low-growing plants. It was built for fast running. It had a short neck, curved back, long tail for balance, and long, slim legs. The arrangement of its toes – four long toes on each front foot and three on the hind feet – also helped give it speed.

OVERLAND TO AUSTRALIA

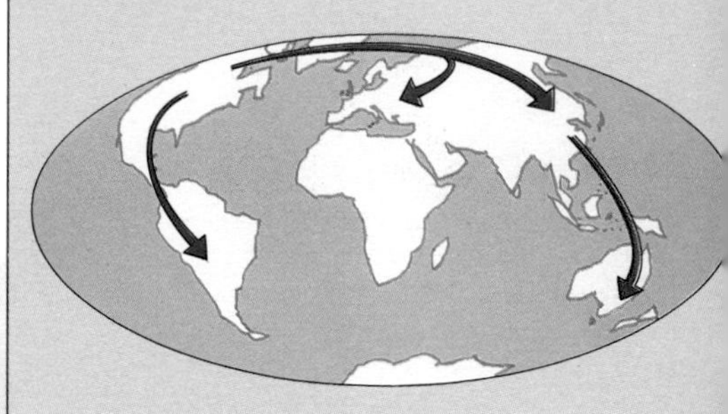

Eurasian route

For a long time scientists have puzzled about how Australia's marsupials got there in the first place. Before the theory of continental drift, geologists thought the continents had always been in their present positions. So the first solution to this puzzle was that marsupials reached Australia by migrating out of North America into northern Europe. Then they must have moved on to Asia before finally arriving in Australia.

Riverbanks and seacoasts were home to many new birds. Ducks, herons, pelicans, and gulls were common.

Moles, camels, rabbits, and voles also evolved, and toward the end of the Eocene, the earliest cats, dogs, and bears appeared. One gigantic bearlike animal called *Andrewsarchus* (actually a hoofed carnivore) had a skull nearly 3 feet long and was easily able to feed on even the biggest herbivores.

Uintatherium was an amblypod as big as a rhinoceros that browsed on soft leaves. It had a six-horned head formed by three pairs of bony knobs. Males also had large tusks, which they used for fighting or defense.

Icaronycteris was the first bat. It looked like a modern bat with its wings formed from tight skin stretched over long, thin fingers. It was insectivorous and probably caught its food at night when it was too dark for other animals to fly safely.

Fossil remains of *Megistotherium* have been found in Eocene rocks in North Africa. It was probably the biggest ever meat-eating land mammal – a huge creodont weighing nearly 2,200 pounds. Its head was twice as big as a grizzly bear's. It was

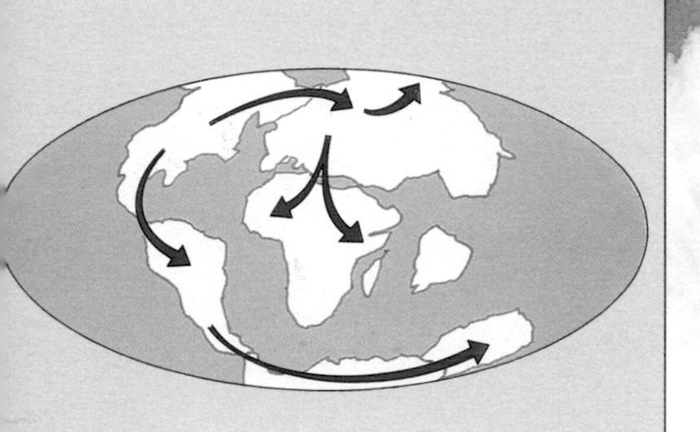

Antarctic route

Continental drift changed scientists' ideas. At the end of the Cretaceous, South America, Antarctica, and Australia lay close together. It was about this time that marsupials started to move into South America from the north. Later, in the Eocene, before Antarctica and Australia separated, marsupial travelers crossed over the Antarctic land bridge to Australia by the southern route. The discovery of marsupial fossils in Antarctica now proves that this hypothesis is correct.

THE FIRST ELEPHANT REPORT

The first elephant looked very different from its relatives that we know today. It was a pig-sized, trunkless animal with a flexible snout and forward-jutting incisor teeth. It lived in swampy areas of North Africa in the late Eocene epoch about 40 million years ago. Scientists have named it *Moeritherium*.

◀ Large flightless birds like *Diatryma* ("terror crane") lived in areas where there were no large meat-eating mammals. They were the Eocene equivalent of today's big predatory cats.

certainly powerful enough to attack some of the biggest mastodonts.

Diatryma was a fast-running "terror crane" from North America that stood nearly 7 feet high. It had a huge parrotlike beak and enormous claws, which it used to kill and tear its prey to pieces. It must have terrorized the early ancestors of the horse, and it is quite likely that *Diatryma* could have "eaten a horse" with no trouble at all! There were several species of giant flightless bird in the Eocene.

The Oligocene Epoch

38 MILLION TO 25 MILLION YEARS AGO

During the Oligocene, climates became cooler as a huge ice cap formed over the South Pole. The formation of such a huge amount of ice used up large amounts of seawater. This caused more dry land to become exposed as the sea level fell around the world. As climates cooled everywhere, the rich tropical forests of the Eocene disappeared in many parts of the world. They were now replaced by woodland, which preferred milder (cooler) climates, and by great areas of grassland.

570 550 525 500 475 450 425 400 375 350 325 300 275 250 225 200 175 150 125 100 75 50 25 00

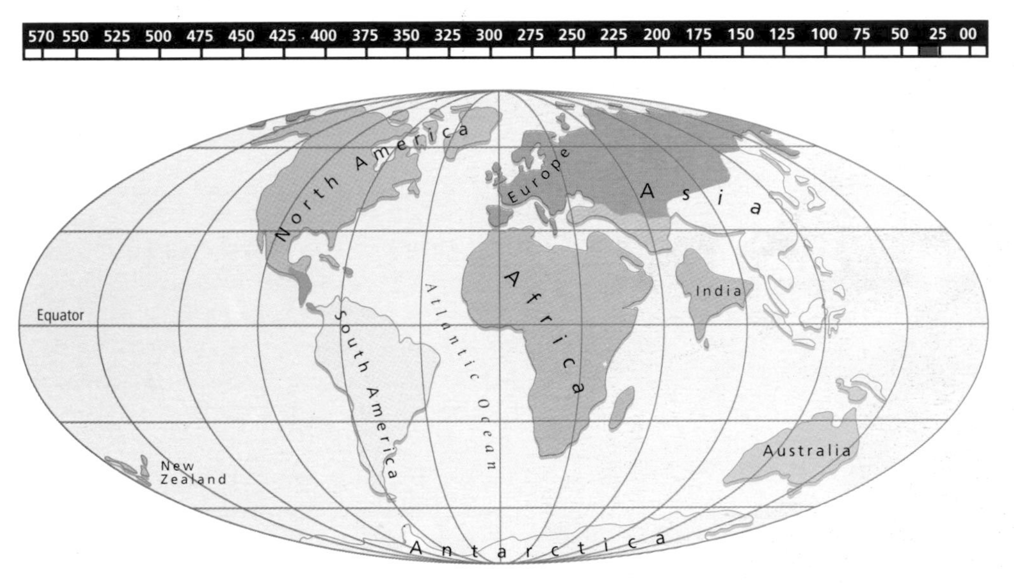

During this period, India crossed north of the equator to lie close to Asia. Australia and Antarctica finally parted company. As Australia drifted away, it carried its population of curious marsupials with it. Now that South America was an island continent, its unusual mammals were also able to evolve in isolation, producing a very strange "zoo" of odd creatures.

More first timers

As grasslands began to spread around the world, herbivores increased in numbers to take advantage of this enormous new food supply. New mammals such as rhinoceroses came on the scene, followed by the first true pigs, cattle, and deer.

Grazing animals have a problem with digestion because grass is difficult to break down. It is not surprising, therefore, that new types of digestive systems were being tried out that could cope with a continuous diet of grass. One of the earliest of these was a design thathas now become the most successful demolisher of cellulose (the building material for plant cell walls) – the ruminant stomach. One of the earliest camels, *Poebrotherium*,

CELLULOSE CRUSHER

Animals such as bison (below), deer, cattle, sheep, and goats, which eat large quantities of grass (mainly cellulose), need special apparatus to digest it. Their four-chambered stomach is designed for the slow digestion of cellulose. The actual breakdown of cellulose takes place in the first chamber, called the rumen. It is performed by millions of bacteria that live there. But it does not happen all at once. A grass meal is continually regurgitated ("vomited up") for more chewing before being swallowed for further digestion in the rumen. The food is passed on to other parts of the stomach only when all the cellulose has been broken down.

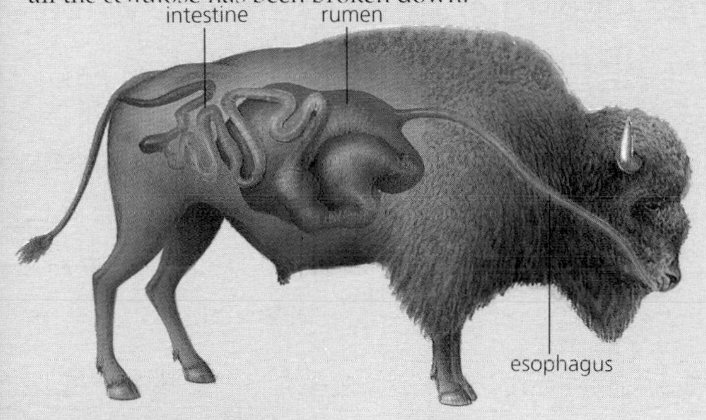

One of the strangest groups of the Oligocene herbivores was the brontotheres. Some were gigantic and looked like a cross between an elephant and a rhinoceros. They had a peculiar Y-shaped bone on the end of their noses that may have helped in driving off attacking predators or in jousting contests between rival males.

Baluchitherium, a gigantic, hornless rhinoceros-type animal, was the biggest land mammal that has ever lived. It stood over 16 feet at the shoulder, measured 26 feet in length, and weighed as much as 17 tons – 8 times the heaviest living rhinoceros. Its enormous height allowed it to browse on leaves as high as a second-story window.

lived at this time. It was one of the first ruminants.

On the track of the herbivores

Just as grass provided a new food source for exploitation by new types of herbivores, so the grass eaters themselves were food for new species of meat eaters. So, toward the end of the Oligocene, the first true cats and dogs appeared.

Strange goings-on in South America

The mammals of South America evolved into all sorts of shapes and sizes once it was cut off from the rest of the world. Many evolved to look remarkably like mammals such as rodents, horses, and elephants that evolved in other parts of the world. *Pyrotherium* resembled an early elephant with its half trunk and chisel-like tusks. *Thylacosmilus* was a large marsupial that looked like a saber-toothed cat. It had long, curved canine teeth and powerful claws. There were also a host of "toothless" mammals called edentates that included the ancestors of anteaters, armadillos, and sloths. Later, even stranger mammals appeared. Then, in the Pliocene, South America became reconnected to the north via a land bridge. After this, things changed once more.

Convergent evolution or old bottles for new wine?

South America is noted for the convergent evolution that took place more than 30 million years ago and that resulted in many of its marsupials gradually evolving to look like placental mammals living in other parts of the world. But South America is by no means unique in this. Throughout evolutionary history, there have been cases where animals living on one continent have evolved to look like animals living on another. Convergent

THE RENEWABLE PANTRY

About 25 million years ago, a new group of plants first appeared that had flowers which were generally small and pale, and which depended on wind for pollination rather than an insect courier system. This group (the grasses) quickly established itself during the Oligocene and Miocene, and soon great areas of the world were covered by grassy plains.

Grass is unique because new leaves grow from the base of the stem rather than the tip, as is the case with other plants. When leaves are cropped by grazing animals, new ones quickly sprout up to take their place, and thus a new meal soon becomes available for the next herd that comes along. This process of continual renewal of food means that grassy plains are capable of supporting large herds of herbivores. The evolution of the grasses provided a new and bigger food source for the plant-eating mammals that evolved later in the Miocene and Pliocene epochs. From this time onward, many different grazers evolved, and because they were an easy target out in the open plains, new predators evolved to hunt them.

► An Oligocene grazing scene would have looked similar to this modern photograph of alpaca grazing on pampas grass in South America. The ancestors of these alpacas would have first appeared in the Oligocene. Grass quickly established itself from the Oligocene because it is so hardy and able to survive conditions which other plants cannot tolerate.

THE SECOND HORSE REPORT

By now, horses had become bigger. *Mesohippus* was the latest model. It was 23 inches at the shoulder, with a straighter back, longer legs, and larger premolar teeth than its Eocene ancestor. Its feet were also different. Each front foot had lost one of its toes and was now a three-toed structure – a design for faster running. Later in the Oligocene, further evolution resulted in the appearance of *Miohippus*. This was bigger than *Mesohippus*, with a big middle toe on each front foot. This arrangement helped lift the body farther off the ground. *Miohippus* could now run even more easily on its toes.

THE SECOND ELEPHANT REPORT

A new type of elephant was now roaming around North Africa. *Phiomia*, an early mastodont ("nipple-toothed") elephant, was bigger than its pig-sized ancestor of the late Eocene. Apart from its extra size (it stood 8 feet at the shoulder) the body shape was also beginning to look more like a modern elephant. Even so, the trunk was still very short. *Phiomia* had an extra-long lower jaw and four short tusks for rooting up low-growing land plants.

evolution takes place when unrelated animals from different parts of the world evolve similar life-styles. They also often develop the same feeding strategy and occupy a similar feeding niche.

There are many examples of convergence. For example, reptilian ichthyosaurs and mammalian dolphins developed similar, streamlined body shapes for cutting through water at high speed. Flying animals such as pterosaurs (reptiles) and bats (mammals) show convergence in the way their wings are designed, each made of a tough membrane of skin stretched tightly over thin support structures made of bone.

Design limitations

At first sight, evolution seems remarkably inventive in the way it seems to arrive at the right answer to a problem to do with survival. Even so, scientists think that there is only a limited number of designs that can properly satisfy a particular need. Natural selection seems to come up with the same kind of answer to a particular design problem over and over again. Take the problem, for example, of designing a structure to slow an animal's rate of descent in free fall. Evolution has begun from many different starting points and moved along numerous paths to finally converge on the idea of a kind of skin "parachute" to solve this specific problem.

Parachutes through the ages

Nearly 230 million years ago, a lizard called *Weigeltisaurus* used a skin parachute stretched between enormously long ribs to glide from tree to tree. *Draco*, a lizard from Southeast Asia, uses the same kind of gliding apparatus today. Flying squirrels plane down (glide) on a membrane on each side of the body that is stretched between the front and back legs. The flying lemur (colugo) is covered from neck to tail by a soft cloak of furred skin that unfolds into the biggest "parachute" in the animal world.

Many plants have also come up with the parachute design to disperse their fruits and seeds. Sometimes this is in the form of a membranelike "wing," as in the sycamore. Other plants, like the dandelion, have developed a modified structure made up of tufts of hairs.

THE MIOCENE EPOCH

25 MILLION TO 5 MILLION YEARS AGO

The continents were still on the move in the Miocene period and some gigantic collisions took place when they met. Africa crashed into Europe and Asia, pushing up the Alps. The Himalayas were squeezed up as India and Asia crunched together. The Rockies and Andes were also formed as other great plates continued to shift and slide over one another. But Australia and South America still remained isolated from the rest of the world, each with its own unique collection of animals and plants.

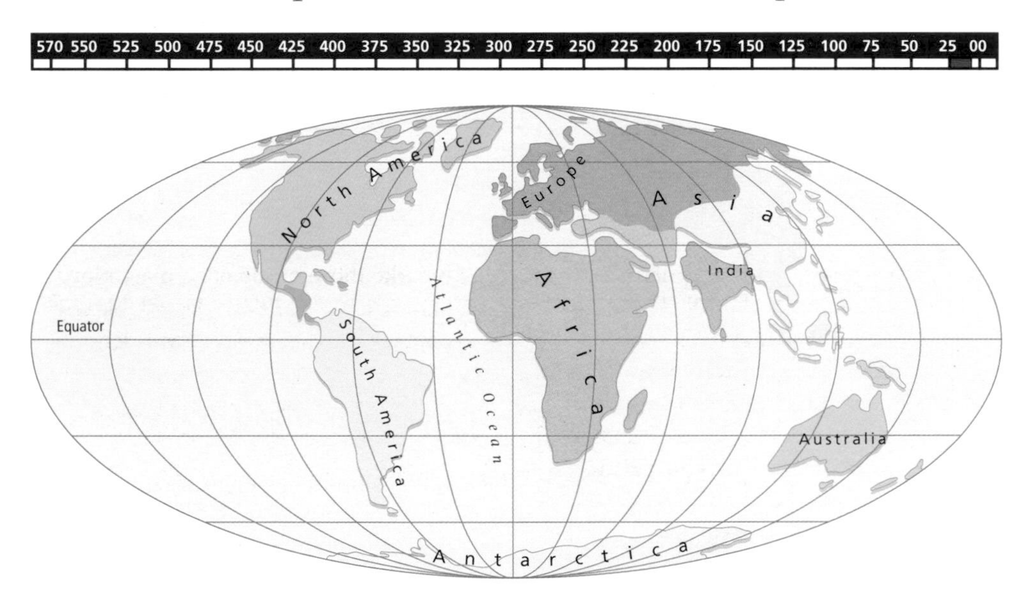

The ice cap that had started to form in the Oligocene covered all of Antarctica during the Miocene. This made climates even cooler than before. As temperatures dropped, grasslands increased to cover more of Africa, Asia, Europe, and North and South America.

More and more herbivores

In the Miocene, there were now many more mammals around, including large numbers of herbivores. By now, the ruminant stomach had evolved into the perfect apparatus for digesting grass. So the Miocene saw an "explosion" in new typesof herbivores that could "chew the cud." Ruminants are able to stuff themselves with enormous amounts of food they can break down later. So a ruminant suddenly attacked by predators can run away with several meal fulls on board. These can be digested at leisure, once the animal has reached a safe place.

As the ruminant revolution took off, ancestors of antelope, cattle, deer, giraffes, and sheep increased in numbers. Antelopelike pronghorns with bizarre

THE THIRD HORSE REPORT

Horses continued to increase in size during the Miocene. *Merychippus* was a pony-sized grazer. The middle toe on each foot was extra-large compared to earlier forms, so it lived its life on tiptoe – an adaptation for fast running. Its cheek teeth were also covered in ridges – an adaptation for chewing tough grass. Previous horses were forest dwellers that fed on soft, lush leaves. But by the Miocene, horses had adapted to a life on the open plains.

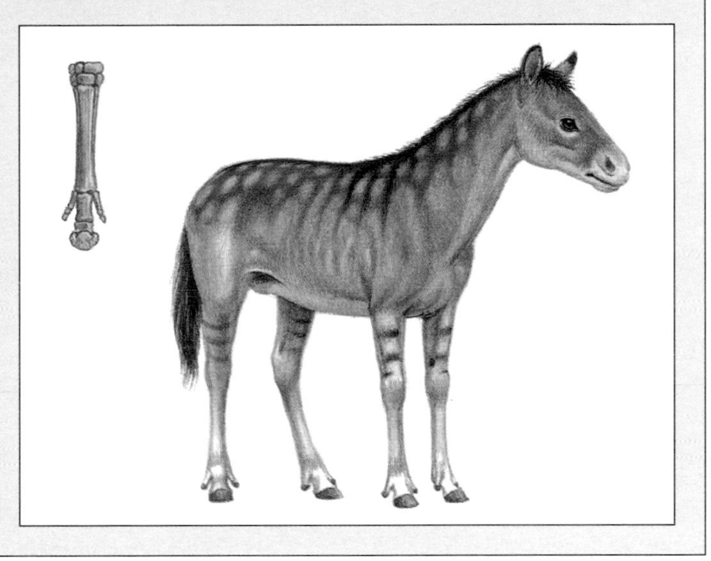

THE THIRD ELEPHANT REPORT

Elephants now looked more like those of today. A mastodon called *Platybelodon* ("shovel tusker") shuffled around like a bulldozer. It had wide, spadelike teeth sticking out of its lower jaw, which it used to scoop plants from the soil. A much bigger elephant called *Deinotherium* (about 13 feet at the shoulder) had curved teeth in the lower jaw, which it may have used like a large fork for digging up roots.

horns on the ends of their noses lived in North America. There were also many new hoofed grazers and browsers, including horses, rhinoceroses, and camels. In Europe and Asia, early types of deer and giraffes had appeared. Africa now had mastodons, the first apes, and monkeys. Lurking in the background were powerful new predators, such as bear-dogs and large saber-toothed cats.

Woodlands support limited numbers of animals

The spread of grasslands and the continued disappearance of forests during the Miocene had a great effect on the increasing populations of herbivorous mammals. It had to do with available energy. Think about a tree for a moment. It uses most of its energy in providing support for itself. It develops a special trunk and many branches to do this. In any year, in a temperate climate, only a small amount of material produced by a tree can be used by animals for food. In deciduous trees (which shed their leaves each year), the leaves are available for only about six months of the year anyway, and the fruits and seeds (apart from nuts) are available for only a few weeks. Because of this, a temperate forest can support only a limited amount of animal life throughout the year.

Grass – the efficient food maker

With grasses, however, the situation is very different. Grasses are generally fast-growing plants that are highly adaptable to climatic change. Most species are low growing – few types grow to more than 3 to 6 feet in height and most are much shorter than this. So a grass plant uses very little energy in building supporting structures because it does not need them. This means that almost the whole plant can concentrate on making and storing food. Grasslands, therefore, form vast areas of photosynthetic activity and, because of this, offer huge food stores for animals to feed on.

New teeth for a tough job

Although the increase in the amount of grass over the earth's surface in the Miocene meant a whole new food source was available, it was not easy at first for herbivorous mammals to exploit this massive new pantry. Some species still became extinct because they were unable to adapt to a grassy diet. It was not easy for mammals whose teeth were designed for eating soft leaves to obtain their food from tough, fibrous grass. For these animals, a diet of grass meant lots of grinding, and this caused their teeth to wear down quickly, something that presented real problems. Mammals do not have an inexhaustible supply of teeth, and a pair of toothless jaws means starvation.

In order to cope with a grass diet, the old-style teeth were redesigned in two ways. Firstly, they became self-sharpening as the pattern of bumps on the chewing surface of the teeth became more intricate. Now, as the teeth gradually wore down, a series of resistant, enamel ridges formed that remained sharp throughout an animal's lifetime. Secondly, individual teeth became bigger because they developed a much larger crown (the part of the tooth above the level of the gum). This, together with the development of open roots (roots with holes in them that allow blood to reach the growing part of the tooth), allowed the teeth to grow throughout an animal's life, so they were now no longer worn away by continually grinding together.

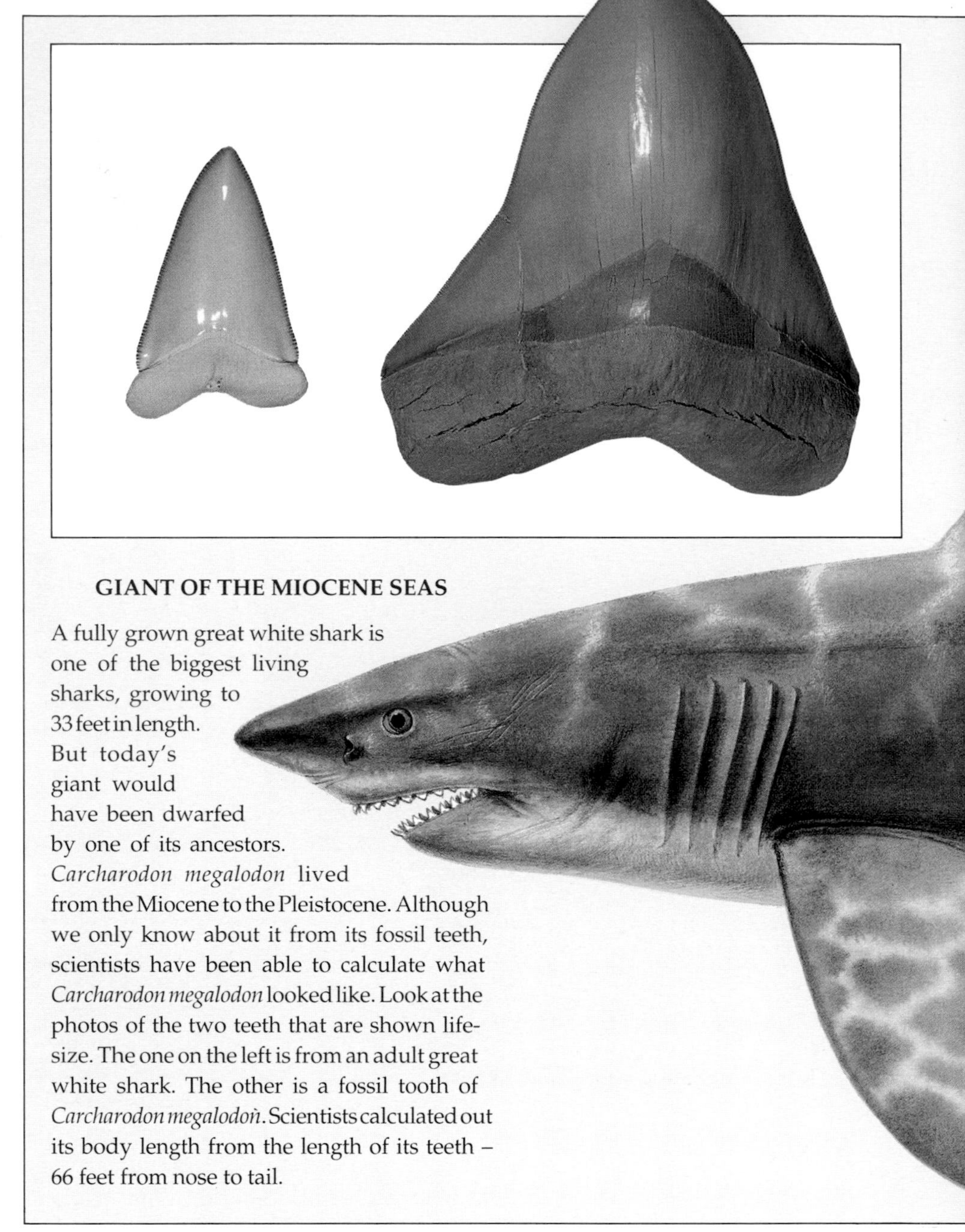

GIANT OF THE MIOCENE SEAS

A fully grown great white shark is one of the biggest living sharks, growing to 33 feet in length. But today's giant would have been dwarfed by one of its ancestors. *Carcharodon megalodon* lived from the Miocene to the Pleistocene. Although we only know about it from its fossil teeth, scientists have been able to calculate what *Carcharodon megalodon* looked like. Look at the photos of the two teeth that are shown life-size. The one on the left is from an adult great white shark. The other is a fossil tooth of *Carcharodon megalodon*. Scientists calculated out its body length from the length of its teeth – 66 feet from nose to tail.

All-around viewing

Being able to deal with a continual supply of grass was only one problem for the newly evolving Miocene herbivores. Living on open grassland and being easily visible to predators was another. To survive in this type of environment, grazers needed good all-around vision – wide-angle or "wraparound" vision (where the eyes are placed on the sides of the head so they can see forward and backward without having to move). This allowed them to see danger approaching from any angle, and so improved the way the herd was able to work together. Cross-herd lookout patterns (individual animals looking diagonally across the herd) and more advanced signaling and communication systems evolved to help increase the chances of survival on the open plains.

Long legs for fast getaways

Long legs are also an advantage for life on the plains. They let an animal stand well above the ground to get a better view of its surroundings. Long legs also provide their owner with the means to escape from danger when necessary. So, during the Miocene period, the limbs of herbivore animals became specialized for speed. The bones in the lower part of the leg gradually became longer while the upper leg bones became shorter. The main muscles used for moving the legs also became shorter and were positioned high up the limbs near where they joined with the body at the shoulder and hip bones.

This kind of arrangement means that an animal is able to take long strides with the minimum use of energy. A grazing animal's legs are lightweight structures and not very strong. But, even so, they are the perfect equipment for a fast getaway and for cruising at high speed over long distances once an animal has got into the rhythm of its run.

The Miocene ecosystem

We can get some idea of what a Miocene ecosystem must have been like by looking at a living example from today's world – the East African savannah. An area of grassland provides a variety of different types of food for animals with the right kind of feeding devices to exploit it. On the savannah of East Africa, zebras eat the coarse tops of the grass while wildebeests and topis eat the leafy central bits. Gazelles find the high-protein seeds and shoots at ground level. The warthog often goes down on its knees to graze on the shortest grass and to root for bulbs and tubers underground. Then there are the browsers that feed above the level of the tallest grass. The black rhinoceros feeds on bark, twigs, and leaves while the elephant is both grazer and browser, often eating up to 550 pounds of vegetation in a single day. The giraffe's great height enables it to avoid competition altogether by collecting twigs and leaves 20 feet above the ground. In this way, the different kinds of grazers and browsers avoid competing with one another, so there is plenty of food to go around. It was probably like this in Miocene times. Different species exploited different parts of the ecosystem.

Other arrivals and new-age travelers

There were other new arrivals. At the beginning of the Miocene, new birds appeared, including parrots, pelicans, pigeons, and woodpeckers. Later, they were joined by the earliest crows and falcons. New mammals such as mice, rats, guinea pigs, and porcupines evolved further. A

▼ The chalicotheres were a strange group of mammals. They looked like a cross between a horse and a rhinoceros. The claws on their feet indicate they were diggers rather than grazers.

strange group of horselike animals called chalicotheres also appeared. They had large claws, rather like hooves, which they used to dig up roots.

Animals were now able to travel easily between Africa, Europe, and Asia. Soon a two-way traffic built up with elephants migrating one way into Eurasia and North America, and cats, cattle, giraffes, and pigs going in the opposite direction.

The dawn of the apes

The earliest primates were small, shrewlike animals that first evolved about 65 million years ago. This line of evolution continued, and by the middle of the Oligocene two main groups were established: the New World monkeys (in South America) and the Old World monkeys (in Africa and Asia). Soon after this, another group evolved from the African branch, which gave rise to the apes, including humans. Apes have larger brains than monkeys. They are also tailless, with long, powerful arms for climbing and swinging through trees.

Scientists have discovered fossils of a small ape called *Aegyptopithecus* ("Egypt ape"), which lived in Africa in the Oligocene about 27 million years ago. No one is sure if this was the original ancestor of modern apes, but it might have been. Soon after the beginning of the Miocene (about 24 million years ago) another, more advanced ape appeared. *Dryopithecus* was a chimpanzee-like ape. It started out in Africa and quickly spread across the land bridges into Europe and Asia. It probably stood on two legs but used all fours for running and climbing. It may even have carried food in its arms. The human story was now about to unfold.

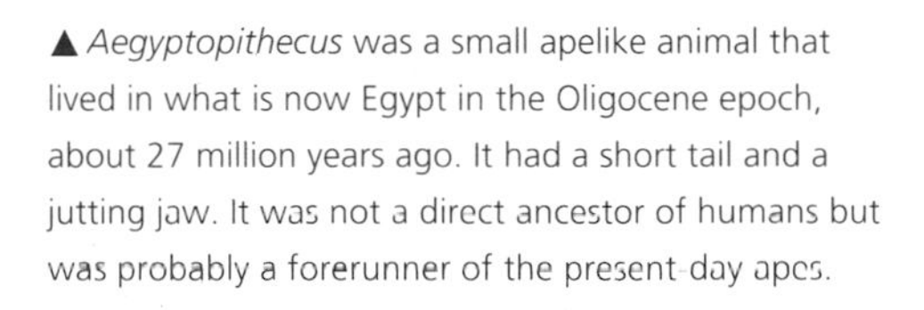

▲ *Aegyptopithecus* was a small apelike animal that lived in what is now Egypt in the Oligocene epoch, about 27 million years ago. It had a short tail and a jutting jaw. It was not a direct ancestor of humans but was probably a forerunner of the present day apes.

◀ African and Asian monkeys (left) have different-shaped noses compared to their South American cousins. Monkeys from the New World (the Americas) have flat noses with widely spaced nostrils opening to the side (far left). Monkeys in the rest of the world have thin noses with forward- or downward-pointing nostrils.

The Pliocene Epoch

5 MILLION TO 2 MILLION YEARS AGO

A traveler from outer space, looking earthward at the beginning of the Pliocene, would have seen the continents positioned almost as they are today. The galactic voyager would also have seen gigantic ice caps in the Northern Hemisphere, in addition to the great ice cap over Antarctica. All this extra ice made the world's climate even cooler, and the earth's landmasses and oceans became much colder. Most of the forests remaining from the Miocene disappeared as grasslands spread worldwide.

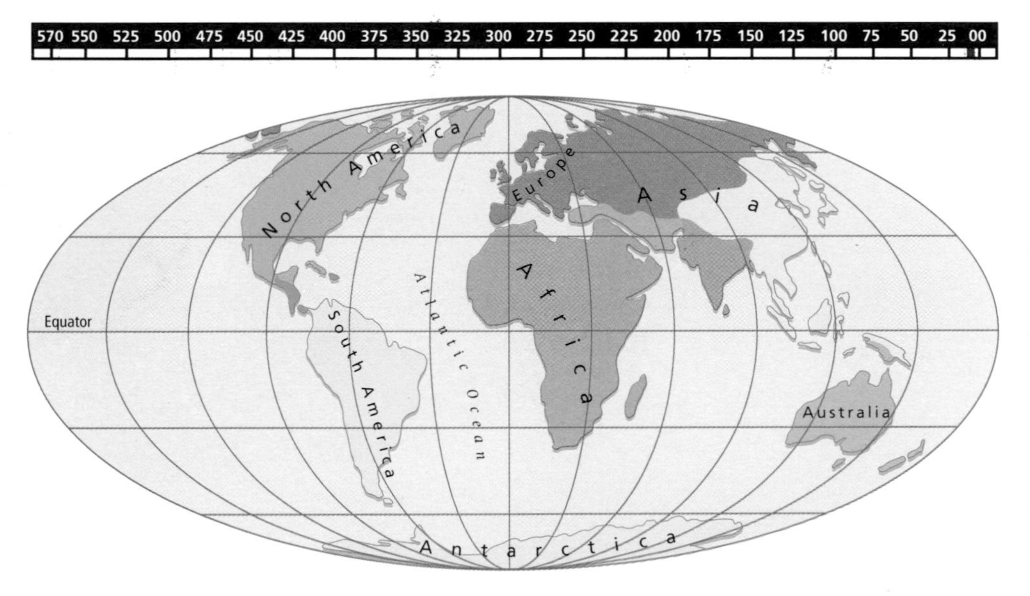

In the Pliocene, grass was still providing an enormous food supply for grazing animals. More advanced ruminants appeared as many leaf browsers died out. In Europe and Asia, cattle, deer, gazelles, and earlier forms of the antelope roamed the grassy plains in large numbers. Deer, camels, horses, mastodons, and pronghorns lived in huge herds on the prairies of North America. There were also short-necked giraffes grazing among the great herds of herbivores. Later, these developed longer necks and became treetop browsers. Hippopotamuses appeared for the first time in the Pliocene. They probably evolved from some kind of piglike ancestor.

The fossil record shows that there were once several species of hippopotamuses. Today there are only two, both living in Africa. Hippopotamuses were probably originally forest-living animals, but as more forests began to be replaced by grasslands in the Pliocene, their ancestors moved out of their forest habitat to live in thickets alongside riverbanks and lake edges. This move brought them close to a vast new food source, and they became grazers on the great grasslands that now existed. From this point on, the evolutionary trend was to become aquatic, nocturnal, and bigger. The common hippopotamus of today spends its daylight hours submerged in water or mud, feeds only at night, and is the second biggest land mammal after the elephant.

More meat eaters

Grazing animals gain protection from predators by living in a herd – there is safety in numbers. Speed also offers a good defense, helping the hunted to escape more easily. As grazing animals developed their herding instincts and speedy running, meat eaters had to adapt. They became more powerful, faster, and extra cunning to catch their food. There were all kinds of cats, dogs, and bears in the Pliocene that preyed on the great herds of plant eaters. In addition to the carnivores that followed the great herds, other smaller meat eaters now made their appearance. Raccoons and weasels attacked smaller prey, and seals chased fish in the cold oceans.

The beginning of a team effort

In the Pliocene, some carnivores developed a new strategy in order to make it easier to catch their prey on the open plains. They began to hunt in packs. This was not an original technique. Dinosaurs such as *Allosaurus* and *Deinonychus* probably adopted a similar way of catching their prey in the Cretaceous, 130 million years previously. But teamwork and co-ordination became more important to Pliocene hunters. Dogs and cats were probably some of the first meat eaters to form hunting teams. One of the advantages about hunting in this way is that an animal much bigger than any one individual pack member can be attacked and killed.

Giant killers

We can see this hunting strategy in use today. The African wild dog hunts in small packs that usually contain no more than seven or eight animals. The victim is chosen before the chase starts and is usually a young member of a grazing herd or a weak or sick-looking individual. The technique is simple. The prey is separated from the

THE FOURTH HORSE REPORT

A horse trial-and-modification experiment had been in progress for 50 million years. *Pliohippus*, the first one-toed horse, had now arrived. The original three-toed design was now obsolete. The side toes had shrunk and finally disappeared. Now a species existed in which each foot had an enlarged middle toe ending in a big, broad hoof. The horse was now properly adapted for life as a swift-footed grazer on the open grassy plains.

▶ One of the most famous carnivores of the late Pliocene is *Smilodon*, the saber-toothed tiger. This leopard-sized predator had long canine teeth in its upper jaw up to 7 inches in length. Scientists used to think *Smilodon* used its huge canines as stabbing weapons, leaping on the back of its prey and piercing through the neck. But stabbing in this way would have probably shattered the canines as they met with bones in the prey's neck or back. It is more likely that *Smilodon* attacked the softer belly or throat, and used its teeth like meat slicers, once it had anchored on.

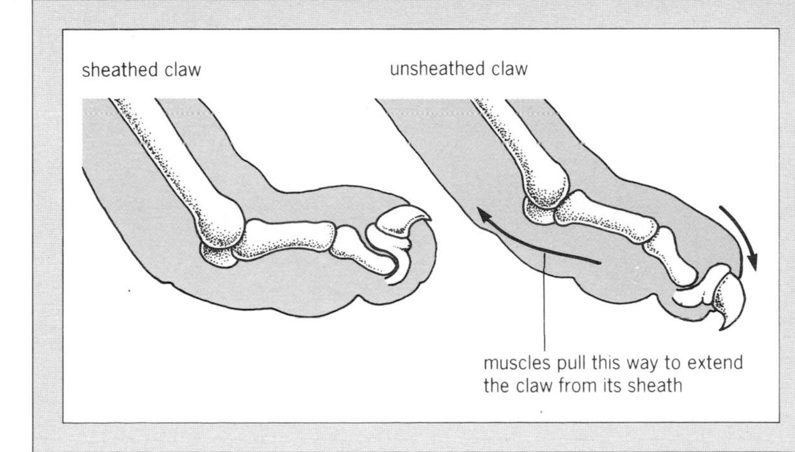

KILLING TOOLS

There are many reasons why cats have become successful hunters. The structure of their skull, their speed, and their well-developed hunting behavior are all important factors. Another is that they can flex (pull out) their claws and spread their toes. This turns each foot into an efficient killing tool. It is all done with muscles and tendons.

rest of the herd and then chased relentlessly until it weakens. When it slows down, the pack moves in and brings it down by individual members grabbing its nose, tail, and belly. The size of the prey varies, but a small team of African wild dogs is capable of killing a zebra 10 times the weight of an individual pack member.

For the good of the pack

Animals that hunt in teams are social creatures, so the pack has to have a proper communal structure based on ability and discipline. Individuals need to know their position within this structure. Good communication among members of the pack is obviously necessary in order to co-ordinate hunting activities. There also has to be a pack leader. Each pack member is an individual with its own personality. However, during a hunt, any individual self-interest must be kept in check for the good of the pack as a whole. The chase is a team effort. It is their pack instinct that has enabled dogs to form such strong relationships with humans and to accept a human being as a leader and trainer.

South America opens up to the world

In the late Miocene and early Pliocene, South America was still home to some unusual "toothless" mammals, the edentates. They included armadillos, tree sloths, and anteaters. There were also grazers like *Toxodon*. With its short legs and broad, three-toed feet, it looked like a rhinoceros. But the position of its nose, eyes, and ears suggests that it lived submerged in water like a hippopotamus.

During the late Pliocene, a narrow land bridge formed that reconnected South America with the rest of the world. A two-way traffic system was set up and a great mammal exchange took place. Tree sloths, anteaters, and *Toxodon* moved into Central America. Opossums and armadillos moved even farther north. Mice, horses, and elephants invaded South America from the north. Now the strange creatures that had lived undisturbed in South America for so long began to face strong competition from the invaders. This, together with changes in climate, made many extinct.

THE FOURTH ELEPHANT REPORT

The descendants of the first elephant experimented with all kinds of food-gathering equipment during their evolutionary development. They tried out jaws in all shapes and sizes. Forks, shovels, spades, and scoops were developed before they finally settled on tusks and a trunk as the best way to get food into the mouth. Elephants also gradually increased in size during the 40 million years up to the beginning of the Pliocene. They were active globe-trotters, invading all continents except Antarctica and Australia. By the Pliocene, mastodons were common. *Stegodon* looked like today's African elephant. It had a long trunk and large, curved tusks. It may have been an ancestor of the giant mammoths that became common about 2 million years later.

◀ *Glyptodon* was a huge armadillo bigger than a car. It looked more like a tortoise than a mammal. It was covered with a great, bony, dome-shaped shell. Its tail was also bony and formed a heavy weapon that it swung like a club when fighting off enemies.

▶ *Megatherium* was a gigantic ground sloth. It was a 20 foot long monster that walked on its knuckles and the sides of its feet. When feeding, it propped itself up on its strong tail, reared up on its back legs, and browsed in the treetops. There is evidence that *Megatherium* existed until fairly recently (10,000 years ago). Some specimens have been discovered with hair still attached to the skin. These parts of the body were probably preserved by the cold dry air in the caves where these remains were discovered.

THE PLEISTOCENE AND HOLOCENE EPOCHS

2 MILLION YEARS AGO TO THE PRESENT

At the beginning of the Pleistocene, most of the continents were in their present-day positions, some of them having moved halfway across the globe to get there. A thin land bridge linked North and South America. Australia was on the other side of the world from Britain. There were huge ice caps in the Northern Hemisphere. The world was in the grip of the Recent Ice Age. The Holocene began 10,000 years ago. The climate became warmer, the glaciers retreated, and humankind flourished.

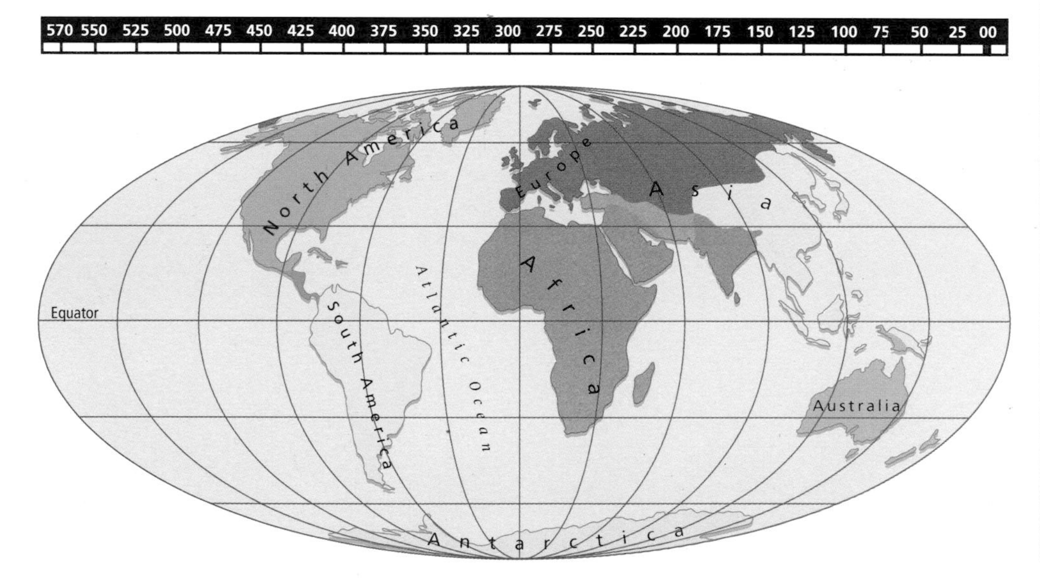

During the Recent Ice Age, the earth froze and then defrosted at least four times. The cold periods were called glacials and the warmer periods interglacials. During a glacial, the ice crept south from the North Pole. Then, in an interglacial, it melted and the ice front moved back again toward the pole. Today we are living in another interglacial – perhaps the fifth in the last 1 million years. During the last glacial, ice caps covered about 11 million square miles of land that are free from ice today. At the same time, the sea level was approximately 492 feet lower than its present-day height.

No one really knows why ice ages occur. Some geologists think that the earth goes through cycles over millions of years, just as we have a yearly cycle of seasons. It may have something to do with the way Earth lies in space or its position in relation to the Sun. Whatever the reason, ice and snow periodically build up at the poles. When this happens, the snow reflects the Sun's heat back into space and Earth becomes colder. This makes more snow and ice form, and Earth cools down even more.

Changing vegetation

Near the front of the creeping glaciers, the land was lifeless. Only a few tiny organisms, called lichens, managed to hold on to the bare rock. Even in regions not covered by ice all year round, the soil became frozen several feet below the surface. In this tundra,

the oak and beech forests that grew before the arrival of the ice now died out. They were replaced by specialized plants, including more lichens and grasses. Huge conifer forests replaced the oak and beech woods.

As the ice then moved back again during an interglacial, temperatures became warmer. Now the specialized cold-climate plants disappeared, and oak and beech forests grew once more. There were also lush grasslands and areas with plenty of flowering plants.

Keeping up with the ice

Many animals died out ahead of the advancing ice. Others migrated south to warmer parts, only to return when the ice retreated. Some mammals adapted to the intense cold by developing thicker body fur that provided better insulation against the freezing temperatures. A time traveler going back to one of the Pleistocene's mini- ice ages in the Northern Hemisphere would have seen a "deep-freeze" world

▲ This modern glacier in Greenland is similar to the huge glaciers that flowed across North America, northern Europe, and the top of Asia during the coldest periods of the Pleistocene. These great rivers of ice carried boulders and rocks embedded in them. Each worked like a gigantic piece of geological sandpaper, scraping and scratching the landscape as it flowed slowly over it. You can see the ancient scratch marks that show where a prehistoric glacier once flowed nearly 2 million years ago.

inhabited by plenty of woolly mammals. There were woolly rhinoceroses and mammoths, extra-hairy reindeer, and shaggy musk-oxen.

The mathematics of body size and climate

A big object has a small surface area compared to its volume, while a small object has a large surface area in relation to its volume. You can easily prove that this is true by figuring out the surface area-to-volume ratios of two different-sized cubes. Try doing this exercise with a 1-inch cube and a 2-inch cube. Calculate the surface area of the 1-inch cube and then figure out its volume. Then, divide the surface area figure by the volume calculation. This gives you the surface area-to-volume ratio of the 1-inch cube. Now repeat the exercise for the 2-inch cube and compare your results.

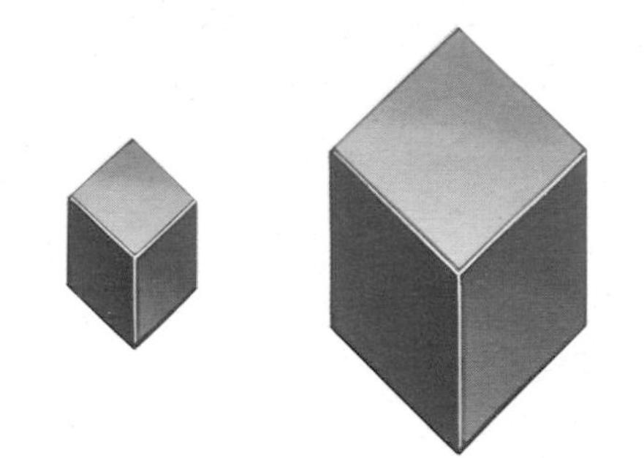

Animals are a little like your cubes. So, for example, a mouse has a bigger surface area-to-volume ratio than an elephant. Mammals are heat generators, but they lose the heat they produce through their skin or body surface. The colder the environment, the more heat they lose and the quicker they cool down. Therefore, a small mammal will lose heat more quickly than a big one. This is because, compared to its volume, it has more body surface to lose it through.

The giant and the dwarf

The woolly mammoth stood nearly 10 feet at the shoulder, and its great bulk and shaggy coat helped conserve body heat so it could survive in cold northern climates. It could even survive beyond the Arctic Circle. Its relative from Sicily was a "dwarf" elephant, less than one-quarter the size of its northern cousin. The small body may have been an adaptation to living on an island. However, being small also helped this little elephant to lose heat and thus keep cool in the hot climate in which it lived. The African elephant seems to break this "rule": It has a huge body but lives in tropical conditions. But remember its large ears! Every time an African elephant flaps its ears, its body surface increases by about 20 percent.

Strange goings-on down under

When Australia was cut off from the rest of the world about 37 million years ago, there were no placental mammals living there. Its only inhabitants were monotremes and marsupials. For a long time after its separation, Australia's resident animals did not have to compete with the up-and-coming placentals, although they did have to face an invasion of rodents and bats later, in the Pliocene. Until late competition arrived, the marsupials were able to evolve in the most unusual ways. In the Pleistocene, there were giant kangaroos 10 feet tall that browsed on trees, and wombat-like animals the size of a hippopotamus. There was even a strange marsupial lion.

▼ We tend to find bigger animals living in cold climates, while their smaller relatives live in warmer parts of the world. The biggest bear is the polar bear from Arctic regions (below left). With a body weight of 1,430 pounds, it is the biggest land carnivore. The sun bear from tropical Southeast Asia (below right) weighs only about one-tenth as much as its polar cousin. Its smaller body adapts it to warm conditions. The polar bear's big body is an adaptation to cold climates.

▲ This deep-frozen baby mammoth was discovered in a glacier in Siberia in 1977. It was so well preserved that its skin and soft tissues were completely intact. Its red blood cells were just like they were when it was alive, and its stomach contents were as fresh as on the day they were eaten 20,000 years ago.

Elephant birds and giant terrible birds

Giant birds were no strangers to the prehistoric world. *Diatryma* (the "terror crane") was a savage hunter 50 million years ago in the Eocene. *Phorusrhacus* also stalked Patagonia in the Miocene, catching prey the size of a goat. Now, in the Pleistocene, even bigger birds existed. The Madagascan elephant bird weighed nearly half a ton and was the heaviest bird of all time. Marco Polo knew about this giant of the bird world and guessed it came from Madagascar. He was right, but actual proof of the world's heaviest bird did not come to light until 1850. And, even then, the evidence was not fossil bones but gigantic eggs, perhaps the biggest eggs ever seen on earth.

In New Zealand there were other giants. Moas were not the heaviest birds of all time but they were certainly the tallest. You could have looked a moa in the eye by gazing at it from an upstairs window. They grew to an enormous height – 11 feet from top to toe. We know a good deal about these gigantic birds because they survived until fairly recently. Scientists have discovered their bones and even the remains of their stomach contents.

As dead as a dodo

The dodo lived on the island of Mauritius in the Indian Ocean. But about 400 years ago, passing sailors started to hunt these docile giants for food. Then the Dutch introduced pigs and monkeys onto Mauritius in the 16th century, and these ate both the dodo's eggs and its young. Now the dodo is no more! Virtually all that is left of this large flightless pigeon are two heads, two feet, and a few skeletons in museums in Europe. But even though no living human has ever seen a dodo, scientists know a good deal about its life-style from studying old ships' logs and the writings of travelers who visited Mauritius before the last bird disappeared 300 years ago.

THE FIFTH HORSE REPORT

After nearly 56 million years of evolution, the modern horse evolved. The Pleistocene version, *Equus*, was a fast-running grazer. It lived in large herds on most continents. Today, Przewalski's horse (a type of wild horse), one species of wild ass, and three types of zebras are the only survivors of these prehistoric herds. The horse has not changed much in the last 2 million years. This means we can get a good idea of what Pleistocene horses looked like by studying their wild descendants alive today.

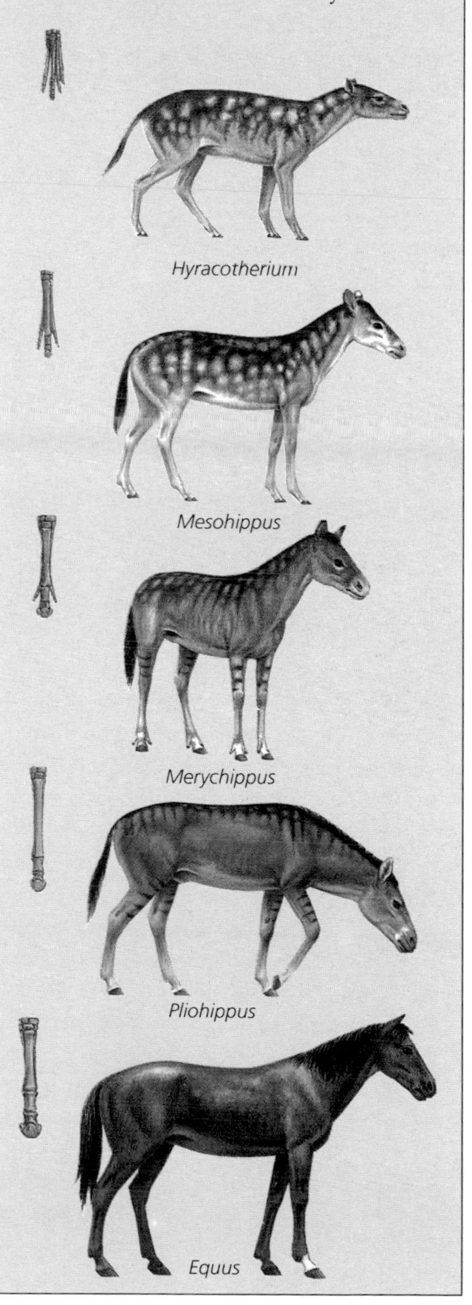

A BRAINTEASER FROM MAURITIUS

The calvaria tree grows on the island of Mauritius. The youngest tree on the island is about 300 years old, and some trees are very much older. Why are none of the calvaria trees on Mauritius younger than 300 years?

One scientist has put forward the theory that the seeds of the calvaria tree needed a dodo to eat them before they could germinate. Like many birds, the dodo had a gizzard (pouch in its gullet) filled with stones to help grind up its food. The seedlings could not germinate until their tough seed cases had been ground open in the dodo's gizzard. However, this link with the dodo has now been questioned by other scientists who think that the seeds of the calvaria tree are able to germinate by themselves.

THE FIFTH ELEPHANT REPORT

The Pleistocene was the "age of the elephants." Over a period of 37 million years, elephants gradually increased in size, developed a long, flexible trunk, and grew huge tusks and grinding teeth. Now, in the Pleistocene, there were many different kinds. There were giants and dwarfs, smooth-coated and woolly species. The biggest-ever elephant was a mammoth that stood 14 feet at the shoulder – 3 feet taller than the biggest African elephant today. At the other extreme, its dwarf cousin was no bigger than a pig.

At the end of the Pleistocene, about 10,000 years ago, the mammoths died out. There are probably many reasons for this. Certainly some could not adapt to warmer climates as the earth came out of the Recent Ice Age. Others were probably killed by humans for food. But, before this happened, modern elephants like those living in Africa and Asia today had already come on the scene.

◄ We know a good deal about the animal life of the Pleistocene because of some unusual fossils preserved in tar pits at Rancho La Brea, near Los Angeles. In this region, prehistoric tar oozed up from the subsoil onto the surface. When heavy rains mixed with the tar, a gluelike mess was produced that formed death traps for any animals that mistook the pools for drinking water. Hundreds of herbivores were caught in these sticky traps over thousands of years. Sometimes, their struggles to escape attracted predators like saber-toothed cats and vultures, which lurked on the edges of the pools or in the sky above, so these animals often became trapped as well. Now this prehistoric graveyard is a wonderful record of life in the Pleistocene nearly 2 million years ago.

Humankind in the Making

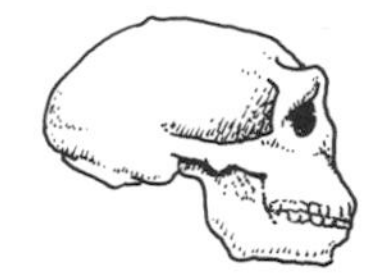

Humans belong to the group of animals called primates. Our earliest ancestors were small, tree-dwelling animals that looked a little like today's tree shrews. They lived about 65 million years ago, at the time when the dinosaurs were dying out. By about 50 million years ago, more advanced types, such as monkeys, appeared. While all this was happening, other primates were evolving in a different way and the first apes appeared about 25 million years ago.

Although today's 180 different species of primates live mostly in tropical or semitropical regions, this has not always been the case. Fifty million years ago, the world was a much warmer place. Then, the ancestors of today's monkeys and apes roamed over a much wider area. Their fossil remains have been found in the British Isles and North America, and as far south as the tip of South America. A chimpanzeelike creature once lived in Europe and Asia. But as the world's climate began to change, the primates living in these regions died out.

Life in the trees

Early primates quickly became skilled tree climbers. Tree climbers need to be able to do two things: They must judge distance and hold on. Forward-facing eyes help the first by giving binocular vision. Grasping fingers are needed for the second. These are two of the most important characteristics of primates. All have movable fingers and thumbs that enable the hands to grip. Some apes and all humans can also bring the tips of the thumb and index finger together to make the letter O. This is the precision grip used for delicate handling. Most important, primates have evolved a big, "thinking" part to the brain to coordinate the activities of eyes and hands.

▼ Today's tree shrews give us some idea of what early primates may have looked like.

The really useful thumbs test

How useful are thumbs? Ask a friend to tape your thumbs across your palms so you cannot use them. Then use one hand only and attempt to pick up things like a pencil or a cup. Try to hold or grip as many things as possible, and make an attempt to eat. You can now see the importance of having an opposable thumb on each hand.

Where we started from

Today there is only one species of human being: *Homo sapiens* (*homo* is Latin for "man" and *sapiens* is Latin for "wise"). But scientists now think there may have been several different species since hominids (human-type animals) first appeared. *Ramapithecus* lived in Africa, Europe, and Asia between 15 and 7 million years ago. It was an apelike animal about 4 feet tall with a flat face and humanlike teeth. It probably spent some

▲ The great apes are our nearest living relatives. Gorillas and chimpanzees live in forested regions of West and East Africa. Gibbons are found in the rain forests of Southeast Asia and orangutans inhabit the steamy jungles of Borneo and Sumatra. Gibbons are less closely related to humans.

of its time hunting in open grassland with sticks and stones. This may have been one of the earliest hominids, but probably not one of our direct ancestors. Scientists now think it is more closely related to the orangutan.

"Southern apes" of Africa

One of the earliest "ape-man" fossils discovered was the skull of a child. It was dug up in 1924 at Taung, in what is now Botswana. The skull had both ape and human characteristics and was called *Australopithecus afarensis*. Since then, many more australopithecine ("southern ape") fossils have been found. They all show that the owners had smallish brains (about 30 cubic inches) and large grinding teeth for eating plants and fruit. They were all very short (about 4 feet tall). Some were big boned and burly, while others were slender and graceful. Some scientists think these were males and females of the same species. Others think they may have been different species. There is much debate about "southern ape," and confusion over the identity of its remains.

The story of Lucy

A remarkable discovery was made when American anthropologist Don Johanson found "Lucy" in Ethiopia in 1974. Lucy was a young female "southern ape" just over 3 feet tall. Her brain and teeth were apelike but she probably walked upright on her bandy legs. Until Lucy's discovery, scientists thought "southern apes" lived about 2 million years ago. But Lucy has been dated at between 3 and 3.6 million years old. This means "southern apes" were around more than 1 million years earlier than scientists first thought.

"Handy man"

While "southern apes" lived in Africa, another group of hominids was beginning to appear and live side by side with them. These were the first true humans, or habilines. They evolved about 2 million years ago, probably from the more slender australopithecines. *Homo habilis* ("handy man") was about the same size as "southern ape" but had a bigger brain – about 43 cubic inches. We know that "handy man" had a tool kit that contained flakes, knifelike implements, choppers, scrapers, cutters, and tools to make more tools.

"Upright man"

Homo erectus ("upright man") appeared about 1.75 million years ago. It was a bigger species than previous hominids (about 5.6 feet tall) and with a much larger brain than "handy man" – 55 cubic inches. At this time, Africa was connected to Europe and Asia. This made it easy for this latest of our ancestors to spread around the world. Fossils have now been discovered in South Africa, Europe, China, and Indonesia. *Homo erectus* made a variety of tools to hunt with, used fire for cooking, and perhaps even developed a simple language. The last of these early humans died out about 150,000 years ago.

▲ "Lucy," the "southern ape" discovered in 1974.

Beijing man gets lost

Beijing man was a type of *Homo erectus* who lived in China about 500,000 years ago. In the 1930s, scientists found a rich collection of fossil bones of this human in a cave near Beijing (Peking). Parts of 45 skeletons were uncovered, including pieces from 14 skulls, 14 lower jaws, 150 teeth, and the bones of 14 children. In 1941, just before war was declared between the United States and Japan, it was decided to send the bones to the United States. This was to stop them from falling into the hands of Japanese soldiers. But the bones never arrived at their destination. They disappeared on their way to a ship that was to take them to safety. To this day, the whereabouts of the bones of Beijing man remain a mystery.

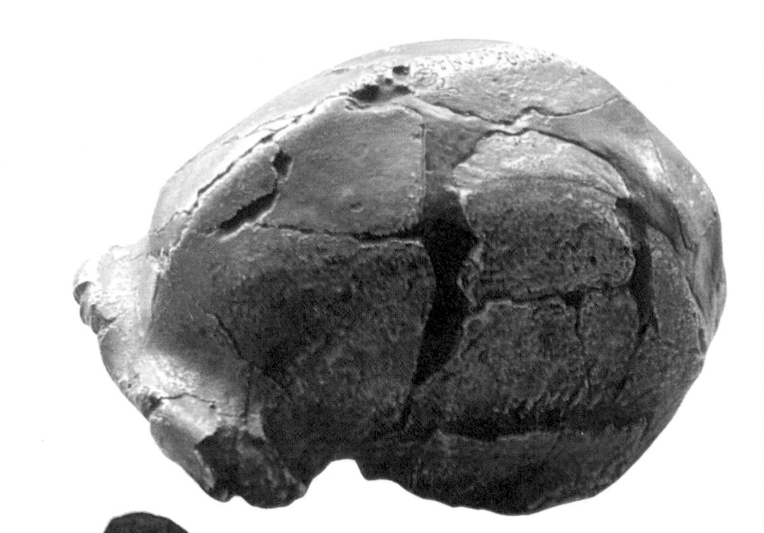

◀ Here are some fragments of bone of the skull of Beijing man – an "upright man." Scientists have been able to put these pieces together to reconstruct a complete skull. It had apelike brow ridges and projecting jaws. The skull also had a bony keel running over the top and a thick ridge on the back. The cranium is bigger than that of *Homo habilis*, and so was the brain.

The Neanderthals

Before "upright man" finally died out, another species of human had started to evolve. *Homo sapiens* ("wise man") first began to appear about 250,000 years ago. Another 180,000 years later (70,000 years ago), Neanderthal humans were living in Europe. Compared to earlier ancestors, Neanderthals were bigger all around, with a large, rounded forehead and a brain as big as ours today – 81 cubic inches. We know a lot about Neanderthals. They lived in the Recent Ice Age, so they wore clothing made from the skin of dead animals and they took shelter in caves. The average life span was about 30 years for men and 23 years for women. They suffered from arthritis, and most of the population was right-handed. There is some evidence that Neanderthals believed in an afterlife. They had funeral ceremonies and even placed flowers in the graves of their dead.

The man who never was

In 1912, some pieces of skull and a broken jawbone were found at Piltdown in Sussex, England. They created great excitement at the time, but some experts soon became puzzled. In 1953, the Piltdown bones were carefully examined and dated. The results were surprising. The jawbone came from a 500-year-old orangutan. The skull was that of a modern human. The bones had been stained and the teeth carefully filed down to make them look prehistoric. The whole thing was a forgery. Piltdown man was a scientific hoax that was finally exposed 40 years after it was carried out. The hoaxer has never been discovered.

Modern humans

Neanderthal humans died out about 30,000 years ago and were replaced by a new, modern type of human. These latest arrivals lived mainly in Africa at first. They were skilled tool makers, good carvers, and excellent painters. As they developed, they began to migrate around the world, and gradually replaced the Neanderthals.

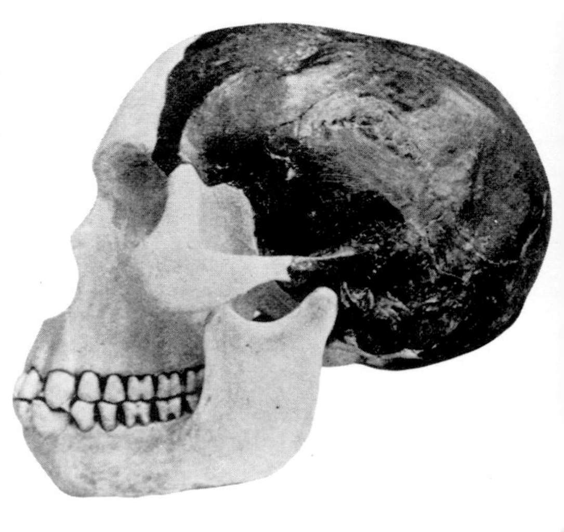

▶ This is a photograph of the Piltdown man skull discovered in Sussex, England, at the beginning of this century. It is now recognized as one of the biggest forgeries in scientific history.

THE HUMAN HUNTERS

Louis Leakey (1903–1972), Mary Leakey (1913–), and their son Richard (1944–) discovered many prehistoric human fossils at Olduvai Gorge in Tanzania. Their first major find was an australopithecine called "nutcracker man." Later, they discovered the first "handy man," and several fossils of "upright man." More recently, Richard Leakey has been excavating in other parts of Africa.

Mary Leakey discovered these remarkable fossil footprints in Tanzania in 1978. They are 3.75 million years old and were made in the mud of volcanic ash that later set solid. They formed a kind of "plaster cast" of our ancestors going for a walk – a family group on a prehistoric outing.

◀ The head of a Neanderthal man.

▼ "Southern apes" may have used stones and bones as tools but habilines were the first to make tools. A stone flake held between fingers and thumb made a good cutting tool. Flatter stones were probably used to scrape meat off bones. Hammer stones were used to make tools with sharp edges. *Homo erectus* invented more advanced tools from pieces of carefully chipped flint. Finer tools were made by Neanderthals. Flint flakes were gouged using pressure with another stone implement and the precision grip.

Modern humans reached Australia about 50,000 years ago and North America only in the last 10,000 years.

As the Recent Ice Age came to an end, modern humans began to change their way of life. They gradually settled down to live in communities. Now the dawn of civilization was about to break. Ten thousand years ago, there were probably about 10 million humans in the world. But about 4,000 years ago their numbers began to increase quickly. By the time Julius Caesar invaded Britain in 55 B.C. the world population had increased to 300 million. Today it is 4 billion and rising.

Getting a "head" start

Recent research has shown that our ancestors probably began walking upright on two legs to stay cool. On the hot African plains, 4 million years ago, walking on two legs gave them several advantages. When upright, the sun's heat would have fallen vertically onto their heads rather than beating down on their backs. Because the top of the head presents a much smaller surface to the sun than an exposed back, our ancestors would have remained cooler. This in turn means they would have sweated less and therefore needed less water to survive. This would have given ancestral humans a "head start" on the road to biological success.

Keeping your hair on

Walking upright has other important implications. For example, a two-legged animal would no longer require the kind of hairy coat needed by other savannah mammals to protect them from the sun's fierce rays beating down on their backs. And so, apart from a hairy scalp that our ancestors kept to shield the part of their body most exposed to the heat of the sun (their head), they became "naked apes."

Keeping your cool

By becoming two-legged walkers, ancestral humans also opened up another important "evolutionary door." In an upright position, much more of an animal's body is raised above the hot ground and away from the heat radiating from it. So, the body and head housing the brain would have been kept cooler than they would at ground level. Additional cooling would have been provided by the breezes that circulate three to six feet above the ground.

When scientists build supercomputers, they have to provide proper cooling systems for them. This is because big computers are very active and produce a lot of heat that must be removed if the computer is to avoid overheating. It is the same with brains. By walking upright, our ancestors lifted up their brains into a cooler environment; and this, combined with their highly efficient cooling system, allowed them to develop bigger, more active brains.

Looking to the future

Human development was a slow process at first. It took nearly 7 million years after our ancestors first appeared for humans to arrive at the stage where they were producing the first cave paintings. But once "wise man" settled down, human talents developed rapidly. Within 100,000 years of making the first cave paintings, humans are the dominant form of life on Earth. We have even managed to escape from our planet altogether by traveling in space.

It is hard to say what humans will look like 10,000 years from now but we will certainly be different. We have changed quite a lot over the past 400 years and even since the beginning of this century. Today's modern soldier would have difficulty fitting into a 15th-century suit of armor. The average height for a medieval soldier was about five feet, five inches. The height of an average soldier today is about five feet, eight inches. Today's fashion model would never be able to squeeze into a dress worn by her great-great-grandmother. Even if she could match her Victorian relative's 18-inch waist, she would be 12 inches too tall! If we continue to evolve in the same way as in the past, our faces will become flatter and our lower jaws smaller. Our brains may be bigger and we will probably grow taller. Since many of us sit down more than we exercise, perhaps our bottoms will become bigger as well!

These drawings show what scientists think our relatives looked like. You can see our ancestors gradually became taller and less apelike as new types evolved.

Australopithecus *Homo habilis*

THE MAN WHO CAME IN FROM THE COLD

On September 19, 1991, a 5,300-year-old man returned to the world. Two hikers in the Austrian Alps found the man's partly exposed body while out for a walk. He was discovered with some clothes and boots, a quiver and two arrows, an ax, a stone-pointed "fire striker," a small flint dagger, a simple knapsack, a sewing kit, and lots of trapping equipment. "Iceman" is the oldest corpse that has ever been found. He lived nearly 1,000 years before the Egyptians started to build the pyramids and 3,000 years before the first Romans.

Homo erectus

Neanderthal human (*Homo sapiens neanderthalensis*)

Modern human (*Homo sapiens sapiens*)

A BRIEF HISTORY OF LIFE ON EARTH

ERA	PERIOD	EPOCH	TIME (millions of years ago)	GEOGRAPHY AND CLIMATE
CENOZOIC ("new life")	Quaternary	Holocene	present–0.01	For the entire Holocene, the continents have been virtually in their present positions, and the climate has been much as today, fluctuating from warmer to colder every few thousand years. Today we are in one of the warmer periods. The sea level has been rising slowly as the ice caps shrink.
		Pleistocene	0.01–2	This was the period of the Recent Ice Age, with fluctuating cold and warmer periods and fluctuating sea levels. This Ice Age continues to the present day.
	Tertiary	Pliocene	2–5	The continents had almost reached their present positions. Huge ice caps spread over the Northern Hemisphere as well as Antarctica and southern South America. The climate became even cooler than in the Miocene.
		Miocene	5–25	Africa crashed into Europe and Asia, pushing up the Alps. India collided with Asia, squeezing up the Himalayas. The Ròckies and Andes also began to form as other continental plates jostled together. The southern ice cap spread to cover all of Antarctica, cooling climates still more.
		Oligocene	25–38	India crossed the equator, and Australia finally separated from Antarctica. The climate cooled and a huge ice cap formed over the South Pole, causing sea levels to fall.
		Eocene	38–55	India moved closer to Asia, and Antarctica and Australia started the epoch close together but began to separate. North America and Europe also separated, and new mountain ranges arose. Seas flooded the land. World climates were warm.
		Paleocene	55–65	The southern continents continued to split up. South America was now completely isolated. Africa, India, and Australia moved even farther apart; Australia remained close to Antarctica. More dry land was exposed and sea levels fell.
MESOZOIC ("middle life")	Cretaceous		65–144	As the continents moved farther apart, the Atlantic Ocean became wider, separating South America and Africa. Africa, India, and Australia, still all south of the equator, began to separate. Large areas of the land were flooded. The remains of shelly planktonic organisms formed great thicknesses of chalk on the ocean floor. World climate was warm and wet at first but cooled later.
	Jurassic		144–213	Pangaea continued to split up, and seas flooded much of the land. There was a lot of mountain building. World climate was warm and dry to start with, then became wetter.
	Triassic		213–248	Pangaea began to split up into Gondwanaland and Laurasia again, and the Atlantic Ocean began to open up. Sea levels worldwide were very low. Temperatures were warm almost all over the world. The climate became gradually drier, causing huge deserts to form inland. Shallow seas and lakes evaporated, becoming very salty.
PALEOZOIC ("ancient life")	Permian		248–286	Gondwanaland and Laurasia moved even closer together, India collided with Asia, and the giant supercontinent of Pangaea was born. The collision threw up mountain ranges. Pangaea began to drift northward. The Permian began with an ice age and falling sea levels. As Gondwanaland moved north, the land warmed and the ice melted. Laurasia became hot and dry, with spreading deserts.
	Carboniferous		286–360	Gondwanaland and Laurasia were moving closer together, forcing up new mountain ranges. In the early Carboniferous there were large areas of shallow coastal sea and swamps and near-tropical conditions over much of the land. The huge luxuriant forests caused a substantial increase in the oxygen content of the atmosphere. Later, the climate cooled, and there were at least two ice ages.
	Devonian		360–408	Gondwanaland in Southern Hemisphere. Laurasia still forming in the tropics. Musch erosion of recently formed mountains, creating large deposits of red sandstones and extensive swampy deltas. The sea level fell toward the end of the period. Climate warmed during the period and became more extreme, with spells of torrential rain and severe droughts. Many parts of the continents became arid.
	Silurian		408–438	Gondwanaland now over the South Pole. The landmasses of North American and Greenland drifted closer together as the Iapetus Ocean shrank. Finally they collided to form the giant supercontinent of Laurasia. A period of great volcanic activity, as new mountain ranges were formed. The period began with an ice age. As the ice melted, sea levels rose and the climate became milder.
	Ordovician		438–500	Gondwanaland still in Southern Hemisphere, with other continents near the equator. Europe and North America were gradually pushed apart by the expanding Iapetus Ocean. As the period progressed, the landmasses moved south. The old Cambrian ice caps melted, raising the sea level. Most of the landmasses were in warm latitudes. A new ice age began at the end of the period.
	Cambrian		500–570	The supercontinent, Gondwanaland, lay across the equator. There were also four smaller continents equivalent to present-day Europe, Siberia, China, and North America. Large stromatolite reefs in shallow tropical waters. Much erosion on land, with large amounts of sediment washed into the sea. Atmospheric oxygen levels rising. An ice age set in toward the end of the period, resulting in a fall in sea levels.
	Precambrian		570–4600	Earth's crust and atmosphere still forming. Later in the Precambrian these early rocks were folded, faulted, metamorphosed, and eroded. In the early Precambrian, Earth was very warm. It has been cooling down ever since. The first recorded ice age occurred around 2.3 billion years ago, and there were two more later in the Precambrian. Between 1 billion and 600 million years ago was the greatest ice age ever known.

ANIMAL LIFE	PLANT LIFE
In the early part of the period, many species became extinct, victims mainly of the warming climate but probably also of increased human hunting. More recently, they could be victims of competition or predation from species of animals introduced into new areas by humans. Human civilizations increased in complexity and have now spread across the world.	Once farming arose, more and more natural vegetation was eliminated to make way for crops and grazing. Also, plants introduced to new areas by humans have sometimes driven out native species.
Some animals adapted to the increasing cold by evolving woolly coats – woolly mammoths and wooly rhinos are examples. Saber-toothed cats and cave lions were the main predators. This was an age of giant marsupials in Australia and of giant flightless birds such as moas and elephant birds in many parts of the Southern Hemisphere. Humans evolved and many large mammals began to disappear.	As the ice spread farther from the poles, tundra and cold grassland replaced the conifer forests, and farther away, conifer forests took over from deciduous woodlands. In warmer parts of the world grasslands flourished.
Grazing hoofed mammals continued to spread and diversify. Toward the end of the period a land bridge linked South and North America, and a great exchange of species took place. It is thought that this new competition drove many species to extinction. Rats arrived in Australia, and the ancestors of humans appeared in Africa.	As the climate cooled, grasslands took over from forests.
Mammals migrated across newly formed land bridges and this stimulated further evolution. Elephants from Africa invaded Eurasia, and cats, giraffes, pigs, and cattle moved in the opposite direction. Saber-toothed cats, monkeys, and apes arose. Monotremes and marsupials continued to diversify in isolation in Australia.	Grasslands spread across the continental interiors as they became cooler and drier.
Herbivores spread and diversified as grasslands expanded, with new species of rabbits, hares, giant sloths, rhinos, and other hoofed mammals. The first ruminants appeared.	The tropical forests shrank and were replaced by temperate woodlands and great areas of grassland. The newly evolved grasses spread rapidly, together with new species of grazing herbivores.
On land, bats, lemurs, tarsiers, and the ancestors of elephants, horses, cattle, pigs, tapirs, rhinos, and deer appeared, as well as other large herbivores. Other mammals, such as whales and sea cows, took to the water. Freshwater bony fish diversified. Other groups were also evolving, including ants and bees, starlings and penguins, giant flightless birds, moles, camels, rabbits and voles, cats, dogs, and bears.	Lush forests grew in many parts of the world, and palms grew in temperate latitudes.
On land, the age of mammals was beginning. Rodents and insectivores evolved, as well as gliding mammals and early primates. Large herbivores and carnivores appeared. At sea, new meat-eating bony fish and sharks took over from the marine reptiles. New forms of bivalves and foraminifers arose.	The flowering plants continued to spread and diversify, along with their pollinators, the insects.
Belemnites became numerous in the seas. Giant turtles and marine reptiles dominated the oceans. On land, snakes evolved, and there were new kinds of dinosaurs and insects such as moths and butterflies. At the end of the period another mass extinction wiped out the ammonites, ichthyosaurs, and many other marine groups, and the dinosaurs and pterosaurs became extinct.	The flowering plants appeared and evolved relationships with insects for pollination. They spread rapidly over the land.
Turtles and crocodiles increased in numbers and variety, and new species of plesiosaurs and ichthyosaurs arose. On land, insects were thriving, including the ancestors of modern ants, bees, caddis flies, earwigs, flies, and wasps. The first bird, *Archaeopteryx,* appeared. Dinosaurs ruled the land, evolving many forms, from the giant sauropods to smaller fast-footed species.	Vegetation spread across the land as the climate became wetter. The ancestors of modern cypresses, pines, and redwoods appeared in the forests.
Dinosaurs and other reptiles became the dominant land animals. The frogs appeared, and later the first tortoises, turtles, and crocodiles. The first mammals appeared, and the mollusks were diversifying. New forms of corals, shrimps, and lobsters evolved. The ammonites almost died out at the end of the period. Marine reptiles such as ichthyosaurs ruled the oceans, while the pterosaurs took to the air.	The cone-bearing plants diversified, forming forests of cycads, monkey-puzzle trees, gingkoes, and conifers. There were also carpets of club mosses and horsetails, and palmlike bennettitaleans.
Bivalve mollusks evolved rapidly. Ammonites became abundant. Modern corals began to take over on reefs. In early Permian amphibians dominated fresh water. Aquatic reptiles evolved, including mesosaurs. In the great extinction at the end of the period, more than 50 percent of animal families disappeared, including many amphibians, ammonites, and trilobites. Reptiles took over from amphibians on land.	The southern landmass was dominated by forests of large *Glossopteris* seed ferns. Conifers appeared and spread inland and up mountains.
Ammonites appeared and brachiopods became more abundant. Rugose, corals, graptolites, trilobites, and some bryozoans, crinoids, and mollusks disappeared. The age of amphibians and also of insects – grasshoppers, cockroaches, silverfish, termites, beetles, and giant dragonflies. In the late Carboniferous the first reptiles appeared.	Dense forsts of giant club mosses, horsetails, tree ferns, and seed plants up to 148 feet tall on deltas and the edges of swamps. The undercomposed remains of these forests developed into coal.
Rapid evolution of fish, including sharks and rays, lobe-finned fish, and ray-finned fish. Ammonites increasing. Giant eurypterids up to seven feet long hunted the seas. In late Devonian, many fish groups became extinct, along with many corals, brachiopods, and ammonites. Many arthropods invaded the land, including mites, spiders, and primitive wingless insects. In late Devonian the first amphibians evolved.	Plants spread from the edges of the water to cover large areas of the land in dense forest. Vascular plants diversified. Spore-bearing lycophytes (club mosses) and horsetails evolved, some developing into trees 125 feet tall.
Rugose corals very active reef builders. Graptolites declining in numbers. Nautiloids, brachiopods, trilobites, and echinoderms thriving. Sea scorpions (eurypterids) in brackish waters. Fish abundant in both salt and fresh water. The first jawed fish, the acanthodians, evolved. Scorpions, millipedes, and possibly eurypterids invaded the land.	Plant life extended around edges of water. Primitive psilopsid plants.
Great increase in filter-feeding animals, e.g. bryozoans (sea mats), sea lilies, brachiopods, bivalve mollusks, and graptolites, which reached their peak in the Ordovician. The archaeocyathids had died out, but stromatoporoids and the first corals arose and carried on reef building. Nautiloids and jawless armored fish increased in numbers.	Algae and seaweeds. First true land plants appeared in late Ordovician.
In a huge burst of evolution, most modern phyla arose, including microscopic foraminifers, sponges, starfish, sea urchins, sea lilies, and velvet worms. Archaeocyathids built huge reefs in the tropics. The first shelled animals appeared; trilobites and brachiopods dominated the seas. The first chordates arose. Later, cephalopod mollusks and primitive fish evolved.	Primitive algae and seaweeds.
The earliest single-celled organisms appeared about 3.5 billion years ago. About 2.8 billion years ago or earlier, the first photosynthesizers (stromatolites) appeared and oxygen levels began to rise. The first multicellular organisms arose about 1.4 billion years ago, and the first cells with nuclei some 1.2 billion years ago. In the late Precambrian came flatworms, segmented worms, jellyfish, and echinoderms.	None

GLOSSARY

algae A group of plantlike photosynthetic organisms, including single-celled organisms and seaweeds.

ammonites An extinct group of cephalopod mollusks with straight or coiled shells. They lived in the sea 395–65 million years ago.

amphibians A group of vertebrates that includes frogs and newts. Amphibians were the first vertebrates to leave water to live on land.

arthropods A group of animals with a hard outer skeleton and jointed limbs. They include insects, spiders, and crustaceans.

australopithecines ("southern apes") A group of pre-human apelike animals that lived in Africa and Asia between 5 and 1 million years ago.

bacteria A group of microscopic single-celled organisms, many of which are involved in decomposing plant and animal remains.

belemnites Squidlike cephalopod mollusks with a bullet-shaped internal shell.

bivalves A group of mollusks, including clams and mussels, whose bodies are protected by a pair of hinged shells.

bony fish A group of fish with a bony skeleton and a single cover over the gills. Most living fish belong to this group.

brachiopods (lampshells) A group of marine animals that extract food from the water using a spiral internal filtering system.

bryozoans (sea mats) Small, colonial animals linked by a hard external skeleton that may form crusts, wavy sheets, or branching structures.

carnivorous Describing a plant or animal that feeds mainly or solely on flesh.

cell The basic unit of which all living organisms are made up. It is a membrane bag that contains living matter and the genetic material of the organism.

cephalopods A group of mollusks, including ammonites, nautiloids, octopus, and squid, in which the front part of the foot forms tentacles for seizing prey.

class In classification, a subdivision of a phylum, containing a number of orders.

club mosses *see* LYCOPHYTES.

coelacanths Large predatory fish whose fins are supported by fleshy lobes. A single species survives today.

cold-blooded Describing an animal (fish, amphibian, or reptile) that cannot control its own blood temperature.

condylarths Medium-sized plant-eating mammals of the Paleocene and Eocene epochs.

conifers Plants such as pines, firs, and spruces, whose seeds are borne in cones.

corals Solitary or colonial animals with a ring of feeding tentacles around the mouth. Reef-building corals secrete a limestone skeleton to cement themselves together.

creodonts The first successful meat-eating placental mammals.

crinoids *see* SEA LILIES.

crustaceans A group of arthropods that includes crabs, shrimps, barnacles, and wood lice.

cycads A group of plants with stout unbranched trunks and long fernlike leaves.

detritus Organic debris formed from decomposing organisms.

dinosaurs A group of reptiles that dominated the land during the Mesozoic era (248–65 million years ago).

echinoderms A group of marine animals, including starfish, sea urchins, and sea lilies, characterized by five-fold symmetry – many of their body parts occur in fives.

elasmosaurs Long-necked marine reptiles of the Cretaceous period.

epoch A subdivision of a geological period.

era A large geological time unit composed of one or more periods.

eurypterids (sea scorpions) An extinct order of predatory crustaceans that lived from Ordovician to Permian times.

evolution The process whereby species originate through modification from earlier forms.

extinction The complete disappearance of a species or group of species.

family In classification, a subdivision of an order, containing one or more genera.

filter feeder An animal that feeds by filtering food particles from the water.

flowering plants Plants that produce flowers and bear seeds enclosed in fruit structures.

foraminifers Microscopic single-celled marine organisms with hard, chalky shells.

fossils The remains, impressions, or traces of organisms preserved in rocks.

fungi A group of non-photosynthesizing, spore-producing organisms that obtain their nutrients by absorbing organic compounds from their surroundings.

gastropods A group of mollusks that includes slugs, snails, and limpets.

genus (plural genera) In classification, a group of species that forms a subdivision of a family.

geological period A division of geological time; for example, the Jurassic period.

geology The study of rocks and the history of Earth.

graptolites A (largely) extinct group of small, colonial aquatic animals. A possible living representative has recently been discovered.

hadrosaurs Plant-eating "duck-billed" dinosaurs; many had elaborate head crests.

herbivorous Plant eating.

horsetails Reedlike plants with hollow stems and whorls of slender branches.

ice ages Periods of low global temperatures when glaciers and ice caps spread over a large part of the earth.

ichthyosaurs A group of dolphinlike aquatic reptiles that lived in the Mesozoic era.

igneous rocks Rocks formed by the solidifying of molten lava or magma.

insects Arthropods whose bodies are divided into three parts – head, thorax, and abdomen – and that have three pairs of legs attached to the thorax.

invertebrates Animals without backbones.

kingdom In classification, the broadest kind of category. The most important kingdoms are Animalia (animals) and Plantae (plants).

lampshells *see* BRACHIOPODS

lava Molten rock pouring from a volcano or from a fissure in the ground.

lycophytes (club mosses) An ancient group of ferns with forking stems covered in small, spirally arranged leaves.

mammals A group of warm-blooded vertebrates that produce milk to feed their young. Humans, whales, and kangaroos are all mammals.

marsupials Mammals such as kangaroos and opossums that give birth to immature young which develop in a pouch on the mother's belly.

metamorphic rocks Rocks formed from preexisting rock that has been changed while deep inside the earth by heat or pressure.

mollusks A group of invertebrates including CEPHALOPODS, GASTROPODS, and BIVALVES.

mosasaurs Giant, fish-eating marine reptiles of the Cretaceous period.

nautiloids A group of cephalopod mollusks that have a spiral shell divided into separate chambers. A few have survived to the present day.

order In classification, a subdivision of a class comprising one or more genera; e.g. Rodentia (mice, squirrels, etc.).

paleontology The study of fossil plants and animals.

pelycosaurs A group of reptiles of the Permian period that had a large sail-like structure on their backs.

period, geological A major unit of geological time; a subdivision of an era.

photosynthesis The process by which plants and various bacteria produce organic compounds, especially sugars, from carbon dioxide and water using sunlight.

phylum In classification, a subdivision of a kingdom comprising one or more classes; e.g., Arthropoda.

placental mammals Mammals such as cats and humans in which the developing young is linked to its mother's blood supply by a pad of tissue (the placenta).

placodonts A group of marine reptiles of the Triassic period that fed on shellfish which they crushed with their flat teeth.

plankton Minute organisms that drift in the sea and fresh water, forming the basis of aquatic food chains.

plesiosaurs An extinct group of large aquatic reptiles that were common in Jurassic and Cretaceous times.

pollen Male reproductive spores produced by seed-producing plants.

predation An interaction between two animals in which one (the predator) attacks, kills, and eats the other (the prey).

prey *see* PREDATION.

pterosaurs A group of flying reptiles that flourished in the Jurassic and Cretaceous periods.

ray-finned fish Bony fish whose fins are stiffened by supports called rays. Most modern fish belong to this group.

reptiles A group of vertebrates including crocodiles, lizards, snakes, turtles, dinosaurs, ichthyosaurs, and pterosaurs.

ruminants A group of mammals, including antelope, cattle, and sheep, that have a special stomach to digest their diet of grass and leaves.

sauropods A group of large, plant-eating dinosaurs of the Jurassic and Cretaceous periods.

scavenger An organism that feeds on dead animal and plant remains.

sea lilies (crinoids) A group of filter-feeding echinoderms with a series of arms surrounding a mouth at the top of a long stalk.

sea mats *see* BRYOZOANS

sedimentary rocks Rocks formed from the weathered remains of preexisting rocks, which have been transported by water, ice, or wind and deposited as beds of sediment.

seed A reproductive structure that consists of a miniature plant, or embryo, and sometimes also a supply of food.

seed ferns An extinct group of fernlike plants that bore seeds on special fronds rather than in cones.

species In classification, a group of organisms that is distinct from other groups of organisms. Members of the same species can interbreed to produce fertile offspring.

sponges Aquatic filter-feeding animals that live attached to some kind of surface. Their soft bodies may be reinforced by silica or calcium carbonate.

stromatolites Round structures built up of layers of cyanobacteria (blue-green algae) and calcium carbonate deposits. They first evolved almost 3 billion years ago.

supercontinent A very large continent formed by the collision of a number of smaller landmasses.

thecodonts "Socket-toothed" reptiles that gave rise to crocodiles, dinosaurs, and pterosaurs.

trilobites A diverse group of arthropods found in Paleozoic seas. The body was typically oval and divided into three regions.

vertebrates Animals that have a backbone made up of units called vertebrae; i.e., fish, amphibians, reptiles, birds, and mammals.

warm-blooded Describing an animal (bird or mammal) that can maintain a constant body temperature.

INDEX

Where several page references are given for a particular word, the more important ones are printed in bold (e.g., **86**). Page numbers in italics (e.g., *94*) refer to illustrations and captions.

JUNIATA COLLEGE
2820 9100 041 683 3
Curr QE 28.3 .B33 1995
Bailey, Jill.
Prehistoric world